tredition

tredition was established in 2006 by Sandra Latusseck and Soenke Schulz. Based in Hamburg, Germany, tredition offers publishing solutions to authors and publishing houses, combined with worldwide distribution of printed and digital book content. tredition is uniquely positioned to enable authors and publishing houses to create books on their own terms and without conventional manufacturing risks.

For more information please visit: www.tredition.com

TREDITION CLASSICS

This book is part of the TREDITION CLASSICS series. The creators of this series are united by passion for literature and driven by the intention of making all public domain books available in printed format again - worldwide. Most TREDITION CLASSICS titles have been out of print and off the bookstore shelves for decades. At tredition we believe that a great book never goes out of style and that its value is eternal. Several mostly non-profit literature projects provide content to tredition. To support their good work, tredition donates a portion of the proceeds from each sold copy. As a reader of a TREDITION CLASSICS book, you support our mission to save many of the amazing works of world literature from oblivion. See all available books at www.tredition.com.

 Project Gutenberg

The content for this book has been graciously provided by Project Gutenberg. Project Gutenberg is a non-profit organization founded by Michael Hart in 1971 at the University of Illinois. The mission of Project Gutenberg is simple: To encourage the creation and distribution of eBooks. Project Gutenberg is the first and largest collection of public domain eBooks.

Twenty-Five Cent Dinners for Families of Six

Juliet Corson

Imprint

This book is part of TREDITION CLASSICS

Author: Juliet Corson
Cover design: Buchgut, Berlin – Germany

Publisher: tredition GmbH, Hamburg - Germany
ISBN: 978-3-8472-1498-4

www.tredition.com
www.tredition.de

Copyright:
The content of this book is sourced from the public domain.

The intention of the TREDITION CLASSICS series is to make world literature in the public domain available in printed format. Literary enthusiasts and organizations, such as Project Gutenberg, worldwide have scanned and digitally edited the original texts. tredition has subsequently formatted and redesigned the content into a modern reading layout. Therefore, we cannot guarantee the exact reproduction of the original format of a particular historic edition. Please also note that no modifications have been made to the spelling, therefore it may differ from the orthography used today.

TWENTY-FIVE CENT DINNERS

FOR

FAMILIES OF SIX.

BY

JULIET CORSON,

Superintendent of the New York Cooking School.

AUTHOR OF "THE COOKING MANUAL,"
"OUR HOUSEHOLD COUNCIL,"
"THE BILL OF FARE, WITH ACCOMPANYING RECEIPTS AND ESTIMATED COST,"
"A TEXT-BOOK FOR COOKING SCHOOLS,"
"FIFTEEN-CENT DINNERS FOR WORKINGMEN'S FAMILIES,"
ETC.

THIRTEENTH EDITION, REVISED AND ENLARGED.

NEW YORK: ORANGE JUDD COMPANY,
245 BROADWAY,
1879.
Copyright by JULIET CORSON, 1878.
All Rights Reserved.

PREFACE

TO

THE REVISED AND ENLARGED EDITION.

During the time that this little book has been a candidate for public favor, it has attained a success far beyond the expectations of its most sanguine advocates; and in issuing this revised and enlarged edition the author returns her sincere thanks to both press and public, who have so substantially seconded her efforts for culinary reform.

In this edition an additional chapter has been devoted to the preparation of fruit for dessert, with special reference to the needs of American housewives. Most American ladies prepare fruit for table use either by canning it, or making it into rich and expensive preserves; while both of these methods are palatable, and available for winter use, the receipts given in the closing chapter will provide a welcome variety for serving fresh fruits at the table, and will tend to increase the healthy consumption of those abundant and excellent domestic productions, while they cannot fail to decrease the deplorable prevalence of that objectionable national compound, the pie.

Recent investigations concerning retail prices in different sections of the country confirm the author in the estimate of cost given in this work; in certain localities some of the articles quoted are more expensive, while others are cheaper; but the average is about equal.
[Pg 3]

PREFACE.

TO ECONOMICAL HOUSEWIVES:

The wide publicity which the press in different sections of the country has given to my offer to show workingpeople earning a dollar and a half, or less, per day, how to get a good dinner for fifteen cents, has brought me a great many letters from those who earn more, and can consequently afford a more extended diet.

In response to their requirements I have written this book, which I hope will be found servicable in that middle department of cookery it is designed to occupy, where we begin to look for more than the absolute necessaries of life; it is a practical guide to the economical, healthful, and palatable preparation of food, and will serve to show that it is possible to live well upon a very moderate income.

It is necessary to repeat in this book some of the directions given in the work on "FIFTEEN CENT DINNERS;" but I hope their reappearance will be pardoned on the ground of their usefulness, and also because the first book will fail to reach many for whom this one is intended.

The cheapest kinds of food are sometimes the most wholesome and strengthening; but in order to obtain all their best qualities we must know how to choose them for their freshness, goodness, and suitability to our needs. That done, we must know how to cook them, so as to make savory and nutritious meals instead of tasteless or sodden messes, the eating whereof sends the man to the liquor shop for consolation.

Good food, properly cooked, gives us good blood, sound bones, healthy brains, strong nerves, and firm flesh, to say nothing of good tempers and kind hearts. These are surely worth a little trouble to secure.

The first food of nearly all living creatures is milk, the only entire natural food; that is, the only food upon which health and strength can be sustained for any length of time, without using any other nourishment. For this reason it is the best food you can give the children if you must restrict their diet at all; and it is also a valuable addition to the food of grown persons. While this fact about milk is settled, it is generally acknowledged by people who study the subject that we [Pg 4] thrive best on a variety. We get warmth and strength from fat meat, wheat, rye, barley, rice, milk, sugar, fruit, peas, beans, lentils, macaroni, and the roots of vegetables; we gain flesh from lean meat, unbolted flour, oatmeal, eggs, cheese, and green vegetables; and, if we want to think clearly, we must use fish, poultry, the different grains, and a good variety of fruit and vegetables.

The food most generally in use among the masses is just that which meets their requirements. No hungry man will spend money for what he knows will not satisfy his appetite, and a natural appetite may always be trusted. For that reason the receipts given in this book treat of the articles in common use, with the exception of lentils and macaroni, which are foods that I earnestly beg all to try. In meals made up of bacon, potatoes and bread, of corned beef and cabbage, and of pork and beans, there exists an equal and sufficient amount of nourishment; but if other dishes are added to these, the variety will result in better general health and contentment. If we were to live day after day on rice, bread, potatoes, or any one other article of food, we would not long be strong enough for any kind of work. In matters of diet variety is not only the spice of life, it is the necessity.

In estimating cost, I have naturally supposed that the family consists of father, mother, and children of different ages, and not of six adults; for them the quantities given would, of course, be insufficient. I allow a meat dinner every day; but in order to have this the meat itself must generally be used one day, with bread or vegetables, and the next day the breakfast must be the broth or juice of the meat, which, if prepared according to my directions, will afford equal nourishment.

I wish to call your attention to the following important fact. The hardy and thrifty working classes of France, the country where the most rigid economy in regard to food is practised, never use tea or coffee for breakfast, and seldom use milk. Their food and drink is BROTH. Not the broth from fresh meat, for they do not often eat that; but that which is made from vegetables, and perhaps a bit of bacon or salt pork.

If you will reflect on the reasons I give in the next chapter for boiling food, instead of roasting or baking it, you will learn two important lessons in economy, namely: that boiling saves at least one fourth the volume of food, and that the broth which is produced, when properly managed, always gives the foundation for another meal. You should always bear in mind that the object of cooking is to soften and disinte [Pg 5] grate food, so that it can be easily masticated; and to expand it, so as to present a large surface to the action of the digestive organs. In this connection you must open your eyes to certain physiological facts if you want your food to agree with you. I shall not tell you more, and perhaps not so much, as you ought to know, and to teach your children.

In calculating the cost of the receipts I give you, I have used the retail prices asked in Washington market, and in ordinary grocery stores, at this season of the year; the average is about the same as that of past years, and probably will not change much; so that I believe I have not placed too low an estimate upon them.

At the first glance it may seem impossible to buy healthy meat at the prices I give, but you must remember that I speak of the good second quality of meat, and that the marketing must be done with economy, and in low-priced localities. It *can* be done, for I have done it myself. Go to packing houses, and provision stores, for meats; to German green-groceries for vegetables, and fruit; and to "speciality" stores, for butter, sugar, tea, et cetera.

In conclusion I only have to say that I hope my little book will be useful to every one who consults it.

JULIET CORSON.

New York Cooking School. [Pg 6]

DAILY BILLS OF FARE FOR ONE WEEK.

MONDAY	Breakfast: Johnny Cake, 5; Cocoa, 6; Broiled Herring, 5.	16	75
	Dinner: Chicken Soup with Rice, 5; Fried Chicken and Potatoes, 35.	40	
	Supper: Tea, 3; Broiled Kidneys, 10; Rice	19	

	Bread, 6.		
TUESDAY	Breakfast: Pulled Bread, 3; Coffee, 6; Macaroni, Farmers' Style, 10.	19	62
	Dinner: Broth and Brewis, 5; Stewed Beef with Norfolk Dumplings, 19.	24	
	Supper: Tea, 3; Peas Pudding, 10; Bread, 6.	19	
WEDNESDAY	Breakfast: Biscuit, 6; Cocoa, 6; Codfish Steaks, 15.	27	76
	Dinner: Spinach Soup, 15; Gammon Dumplings and Potatoes, 15.	30	
	Supper: Tea, 3; Baked Beans, 10; Potato Bread, 6.	19	
THURSDAY	Breakfast: Breakfast Rolls, 6; Cocoa, 6; Stewed Pig's Kidneys, 10.	22	69
	Dinner: Macaroni with White Sauce, 10; Brain and Liver Pudding, with Potatoes, 18.	28	
	Supper: Tea, 3; Rice, Japanese Style, 10; Bread, 6.	19	
FRIDAY	Breakfast: Indian Cakes, 5; Coffee, 6; Vegetable Porridge, 15.	26	66
	Dinner: Thick Pea Soup, 6; Fish and Potato Pudding, 15.	21	
	Supper: Tea, 3; Fried Beans, 10; Rice Bread, 6.	19	
SATURDAY	Breakfast: Biscuit, 6; Cocoa, 6; Rice, Milanaise Style, 10.	22	58
	Dinner: Mutton Broth, with Barley, 3; Epigramme of Lamb, 16; Potatoes, 3.	22	
	Supper: Tea, 3; Polenta, 5; Potato Bread, 6.	14	
SUNDAY	Breakfast: Toast, 6; Fried Lentils, 10; Coffee, 6; Oatmeal Porridge, 8.	30	1.19

	Dinner: Roast Fowl and Baked Potatoes, 38; Half-pay Pudding, 20.	58	
	Supper: German Potatoes, 10; Cream Rice Pudding, 15; Bread, 6.	31	
TOTAL.			$5.25

CHAPTER I.

MARKETING.

The most perfect meats are taken from well-fed, full-grown animals, that [Pg 10] have not been over-worked, under-fed, or hard-driven; the flesh is firm, tender, and well-flavored, and abounds in nutritious elements. On the other hand, the flesh of hard-worked or ill-fed creatures is tough, hard, and tasteless.

All animal flesh is composed of albumen, fibrin, and gelatin, in the proportion of about one fifth of its weight; the balance of its substance is made up of the juice, which consists of water, and those soluble salts and phosphates which are absolutely necessary for the maintenance of health. It is this juice which is extracted from beef in the process of making beef tea; and it is the lack of it in salted meats that makes them such an injurious diet when eaten for any length of time to the exclusion of other food.

The flesh of young animals is less nutritious, and less easily masticated than that of full grown animals, on account of its looser texture. Beef, which has firmer and larger fibres than mutton, is harder to digest on that account, but it contains an excess of strengthening elements that is not approached by any meat, save that of the leg of pork.

The tongues of various animals, the fibres of which are small and tender, are nutritious and digestible; the heart is nutritious because it is composed of solid flesh, but the density of its fibre interferes with its digestibility; the other internal organs are very nutritious, and very useful as food for vigorous persons on that account, and because they are cheap. The blood of animals abounds in nutritive elements; the possibility of its use as a general food has closely engaged the attention of European scientists; notably of the members of the University of Copenhagen, who recommend its use in the following forms, in which it is not only suitable for food, but also capable of preservation for an indefinite time. [Pg 11] First, as sausages, puddings and cakes—being mixed with fat, meal, sugar, salt, and a few spices—to serve as a much cheaper substitute for meat, and intended especially for the use of the poor classes; and second, as blood-chocolate, more especially suitable to be used in hospitals, as well as otherwise in medical practice, in which latter form it has been recommended by Professor Panum, at a meeting of physicians at Copenhagen, and is now being employed in some of the hospitals of that city.

Bones consist largely of animal matter, and earthy substances which are invaluable in building up the frame of the body. In order to obtain all their goodness, we must crush them well before putting them into soups or stews.

Beef.—The flesh of the best quality of beef is of a bright red color, intersected with closely laid veins of yellowish fat; the kidney fat, or suet, is abundant, and there is a thick layer upon the back. The second quality has rather whitish fat, laid moderately thick upon the back, and about the kidneys; the flesh is close-grained, having but few streaks of fat running through it, and is of a pale red color, and covered with a rough, yellowish skin. Poor beef is dark red, gristly, and tough to the touch, with a scanty layer of soft, oily fat. Buy meat as cheap as you can, but be sure it is fresh; slow and long cooking will make tough meat tender, but tainted meat is only fit to throw away. Never use it. You would, by doing so, invite disease to enter the home where smiling health should reign. The best way to detect taint in any kind of meat is to run a sharp, thin-bladed knife close to the bone, and then smell it to see if the odor is sweet. Wipe the knife after you use it. A small, sharp wooden skewer will answer, but it must be scraped every time it is used, or the meat-juice remaining on it will become tainted, and it will be unfit for future use. If, when you are doubtful about a piece of meat, the butcher refuses to let you apply this test carefully enough to avoid injuring the meat, you will be safe in thinking he is afraid of the result.

Mutton.—Prime mutton is bright red, with plenty of hard, white fat. The flesh of the second quality is dark red and close grained, with very few threads of fat running through it; the fat is rather soft, and is laid thin on the back and kidneys, closely adhering to them. The poorest healthy quality has very pale flesh, and thin white fat, and the meat parts easily from the bone. Diseased mutton has decidedly [Pg 12] yellow fat, and very soft flesh, of loose texture. Tainted mutton smells bad; test it as you would beef.

Lamb.—A carcass of lamb should weigh about twenty-five pounds before it is old enough to be wholesome and nourishing food; before it has reached that age it is watery and deficient in the elements of strength; at any age it is more suitable food for women and children than for healthy men. The finest kind has delicate rosy meat, and white, almost transparant fat. The flesh of the second quality is soft, and rather red compared with the pinkish-white meat of choice kinds; the fat is more scanty, and the general appearance coarser. The poorest lamb has yellow fat, and lean, flabby red meat, which keeps but a short time. Test the freshness of lamb by touching the kidney-fat; if it is soft and moist the meat is on the verge of spoiling; a bad smell indicates that it is already tainted; it is utterly unfit for use.

Veal.—Prime veal is light flesh color, and has abundance of hard, white, semi-transparent fat. The flesh of the second quality is red in contrast to the pinkish-white color of the prime sort; and the fat is whiter, coarser-grained, and less abundant. The poorest kind has decidedly red flesh, and very little kidney-fat. The neck is the first part that taints, and it can easily be tested; the loin is just spoiling when the kidney-fat begins to grow soft and clammy.

Read this sentence about BOB-VEAL carefully, and be sure to remember it. It is the flesh of calves killed when two or three weeks old, or that of "deaconed calves," which are killed almost as soon as they are born, for the value of their skins. This practice cannot be too harshly condemned as a criminal waste of food; for a stock raiser, or farmer, who knows his business can feed his calves until they reach a healthy maturity, without seriously interfering with his supply of milk. The flesh of BOB-VEAL is a soft, flabby, sticky substance, of a ropy gelatinous nature; and, being the first flesh, unchanged by the health-giving action of air and food, it is devoid of the elements necessary to transform it into wholesome food. IT SHOULD NEVER BE EATEN.

Pork.—The best kind of pork is fresh and pinkish in color, and the fat is firm and white. The second quality has rather hard, red flesh, and yellowish fat. The poorest kind has dark, coarse grained meat, soft fat, and discoloured kidneys. The flesh of stale pork is moist and clammy, and its smell betrays its condition. Measly pork has little kernels in the fat, and is unhealthy and dangerous food. After testing, [Pg 13] as you would beef, so as to see if it is fresh, and making sure that it is not measly, we have still to dread the presence of TRICHINA, a dangerous parasite present in the flesh of some hogs. The surest preventive of danger from this cause is thorough cooking, which destroys any germs that may exist in the meat. Cook your pork until it is crisp and brown, by a good, steady fire, or in boiling water, at least twenty minutes to each pound. Pork eaten in cold weather, or moderately in summer, alternately with other meats, is a palatable and nutritious food. It has a hard fibre, and needs to be thoroughly chewed in order to be perfectly digested; for that reason it should be sparingly used by the young and the very old. The least fat is found in the leg, which contains an excess of flesh-forming elements, and resembles lean beef in composition; the most fat is in the face and belly. When cured as bacon it readily takes on the anti-septic action of salt and smoke, and becomes a valuable adjunct to vegetable food, as well as a pleasant relish; and in this shape it is one of the most important articles in general use.

Poultry.—Both poultry and game are less nutritious than meat, but they are more digestible, and consequently are better food than meat for persons of weak digestive organs and sedentary habits. They are both excellent for persons who think or write much. Fresh poultry may be known by its full bright eyes, pliable feet, and soft moist skin; the best is plump, fat, and nearly white, and the grain of the flesh is fine. The feet and neck of a young fowl are large in proportion to its size, and the tip of the breast-bone is soft, and easily bent between the fingers; a young cock, has soft, loose spurs, and a long, full, bright red comb; old fowls have long, thin necks and feet, and the flesh on the legs and back has a purplish shade; chickens and fowls are always in season.

Turkeys are good when white and plump, have full breasts and smooth legs, generally black, with soft loose spurs; hen turkeys are smaller, fatter, and plumper, but of inferior flavor; full grown turkeys are the best for boiling, as they do not tear in dressing; old turkeys have long hairs, and the flesh is purplish where it shows under the skin on the legs and back. About March they deteriorate in quality.

Young ducks and geese are plump, with light, semi-transparent fat, soft breast-bone, tender flesh, leg joints which will break by the weight of the bird, fresh colored and brittle beaks, and windpipes that break [Pg 14] when pressed between the thumb and forefinger. They are best in fall and winter.

Young pigeons have light red flesh upon the breast, and full, fresh colored legs; when the legs are thin, and the breast is very dark, the birds are old.

Game Birds.—Fine game birds are always heavy for their size; the flesh of the breast is firm and plump, and the skin clear; and if a few feathers be plucked from the inside of the leg and around the vent, the flesh of freshly killed birds will be fat and fresh colored; if it is dark, and discolored, the game has been hung a long time. The wings of good ducks, geese, pheasants, and woodcock are tender to the touch; the tips of the long wing feathers of partridges are pointed in young birds, and round in old ones. Quail, snipe, and small birds should have full, tender breasts.

Fish.—Fish is richer in flesh-forming elements than game, poultry, lamb or veal, but it contains less fat and gelatin. It is easily digested, and makes strong muscular flesh, but does not greatly increase the quantity of fat in the body. The red blooded and oily kinds, such as salmon, sturgeon, eels and herring, are much more nutritious than the white blooded varieties, such as cod, haddock, and flounders. The salting of rich, oily fish like herring, mackerel, salmon, and sturgeon, does not deprive it of its nutritive elements to the extent that is noticeable with cod; salt cod fish is almost entirely devoid of nutriment, while the first named oily varieties are valuable adjuncts to a vegetable diet.

Although fish contains more water and less solid nutriment than meat, it is generally useful from its abundance and cheapness; and certain kinds which are called red-blooded, are nearly as nourishing as meat: oily fish satisfies hunger as completely as meat; herring, especially, makes the people who eat it largely strong and sinewy. Sea fish are more nourishing than fresh water varieties.

Sea fish, and those which live in both salt and fresh water, such as salmon, shad, and smelts, are the finest flavored; the muddy taste of some fresh water species can be overcome by soaking them in cold water and salt for two hours or more before cooking; all kinds are best just before spawning, the flesh becoming poor and watery after that period. Fresh fish have firm flesh, rigid fins, bright, clear eyes, and ruddy gills. [Pg 15] Oysters, clams, scallops, and mussels, should be eaten very fresh, as they soon lose their flavor after being removed from the shell.

Lobsters and crabs should be chosen by their brightness of color, lively movement, and great weight in proportion to their size; you ought always to buy them alive, and put them head first into a large pot of boiling water, containing a handful of salt; they will cook in about twenty minutes.

Vegetables.—In order to be healthy we must eat some fresh vegetables; they are cheap and nourishing, especially onions and cabbages. Peas, beans, and lentils, all of which are among the lowest priced of foods, are invaluable in the diet of a laboring man: he can get so much nourishment out of them that he hardly needs meat; and if they are cooked in the water that has been used for boiling meat, they make the healthiest kind of a meal.

All juicy vegetables should be very fresh and crisp; and if a little wilted, can be restored by being sprinkled with water and laid in a cool, dark place; all roots and tubers should be pared and laid in cold water an hour or more before using. Green vegetables are best just before they flower; and roots and tubers are prime from their ripening until they begin to sprout.

When it is possible buy your vegetables by the quantity, from the farmers, or market-gardeners, or at the market; you will save more than half. Potatoes now cost at Washington market from one to one dollar and a half a barrel; there are three bushels in a barrel, and thirty-two quarts in a bushel; now at the groceries you pay fifteen cents a half a peck, or four cents a quart; that makes your barrel of potatoes cost you three dollars and sixty-three cents, if you buy half a peck at a time; or three dollars and eighty-four cents if you buy by the quart. So you see if you could buy a barrel at once you could save more than one half of your money. It is worth while to try and save enough to do it.

Fruit.—Fresh fruit is a very important food, especially for children, as it keeps the blood pure, and the bowels regular. Next to grains and seeds, it contains the greatest amount of nutriment to a given quantity. Apples are more wholesome than any other fruit, and plentiful and cheap two-thirds of the time; they nourish, cool, and strengthen the body. In Europe laborers depend largely upon them for nourishment, and if they have plenty, they can do well [Pg 16] without meat. They miss apples much more than potatoes, for they are much more substantial food.

All fruit should be bought ripe and sound; it is poor economy to buy imperfect or decayed kinds, as they are neither satisfactory nor healthy eating; while the mature, full flavored sorts are invaluable as food.

Preserved and dried fruits are luxuries to be indulged in only at festivals or holidays. Nuts are full of nutritious oil, but are generally hard to digest; they do not come under the head of the necessaries of life.

CHAPTER II.

HOW TO COOK, SEASON, AND MEASURE.

Before beginning to give you receipts, I wish to tell you about the effect of cooking food in different ways. We all want it cooked so that we can eat it easily, and get the most strength from it, without wasting any part of it. I will tell you some very good reasons for making soup and stew out of your meat instead of cooking it in any other way.

Roasting or Baking.—The first is the most extravagant way of cooking meat, as it wastes nearly one third of its substance in drippings and steam; the second also is very wasteful, unless the meat is surrounded with vegetables, or covered with a flour paste. When you do bake meat without a covering of paste, put it into a hot oven at the start, to crisp the outside and to keep in the valuable juices; you can moderate the heat of the oven as soon as the meat is brown, and let it finish cooking slowly by the heat of the steam which is constantly forming inside of it. It generally takes twenty minutes to bake each pound of meat.

[Pg 17]

Broiling.—This is another extravagant way of cooking meat, for a great deal of the fat runs into the fire, and some nourishment escapes up the chimney with the steam. If you must broil meat, have your fire hot and clear, and your gridiron perfectly clean; and, unless it has a ledge to hold the drippings, tip it towards the back of the fire, so that the fat will burn there, and not blacken the meat as it would if the gridiron were laid flat, and the fat could burn under the meat. Never stick a fork into broiled meat to turn it; and do not cut it to see if it is done; for if you do either you will let out the juice. Study the following table, and then remember how near the time given in it comes to cooking according to your taste. Fish will broil in from five to ten minutes; birds and poultry in from three to fifteen minutes; chops in from ten to fifteen minutes, and steak in from ten to twenty minutes.

Boiling and Stewing.—Boiling food slowly, or stewing it gently, saves all its goodness. After the pot once boils you cannot make its contents cook any faster if you have fire enough under it to run a steam engine; so save your fuel, and add it to the fire, little by little, only enough at a time to keep the pot boiling. Remember, if you boil meat hard and fast it will be tough and tasteless, and most of its goodness will go up the chimney, or out of the window, with the steam. Boil the meat gently, and keep it covered close to save the steam; it will condense on the inside of the cover, and fall back in drops of moisture upon the meat.

The following table shows how much is wasted in the different ways of cooking we have just spoken of. Four pounds of beef waste in boiling or stewing, about one pound of substance, but you have it all in the broth if you have kept the pot covered tightly; in baking one pound and a quarter is almost entirely lost unless you have plenty of vegetables in the dripping pan to absorb and preserve it; in roasting before the fire you lose nearly one pound and a half. Do not think you save the waste in the shape of drippings; it is poor economy to buy fat at the price of meat merely for the pleasure of trying it out.

Frying.—This is a very good method of cooking fish, and of warming cold meat and vegetables. To fry well put into your frying pan enough fat to cover what you mean to fry, and let it get smoking hot, but do not burn it; then put in your food, and it will not soak fat, and will generally be done by the time it is nicely browned. To SAUTÉ, or HALF FRY any article, you should begin by putting in the pan enough fat to cover the bottom, and let it get smoking hot, but not burnt before you put in the food. This also is a good way to warm over meat, vegetables, oatmeal, or pudding.

A very good way to cook meat and vegetables together is to put them in an earthen jar, cover it tightly, and cement the cover on with flour paste; then bake for about four hours.

If you are going to use a piece of meat cold do not cut it until it [Pg 18] cools, and it will be more juicy. If the meat is salt let it cool in its own pot liquor, for the same reason.

Salt and Smoked Meats.—These meats are best when they are put over the fire in cold water, brought gradually to a boil, and then set back from the fierce heat of the fire, so as to keep scalding hot WITHOUT BOILING; they take longer to cook this way, but they are tender and delicious, and very little fat is wasted.

Seasoning Food.—Many people have the idea that a finely flavored dish must cost a great deal; that is a mistake; if you have untainted meat, or sound vegetables, or even Indian meal, to begin with, you can make it delicious with proper seasoning. One reason why French cooking is so much nicer than any other is that it is seasoned with a great variety of herbs and spices; these cost very little; if you would buy a few cents' worth at a time you would soon have a good assortment. The best kinds are Sage, Thyme, Sweet Marjoram, Tarragon, Mint, Sweet Basil, Parsley, Bay-leaves, Cloves, Mace, Celery-seed, and onions. If you will plant the seed of any of the seven first mentioned in little boxes on your window sill, or in a sunny spot in the yard, you can generally raise all you need. Gather and dry them as follows: parsley and tarragon should be dried in June and July, just before flowering; mint in June and July; thyme, marjoram and savory in July and August; basil and sage in August and September; all herbs should be gathered in the sun-shine, and dried by artificial heat; their flavor is best preserved by keeping them in air-tight tin cans, or in tightly corked glass bottles.

Dried Celery and Parsley.—If you ever use celery, wash the leaves, stalks, roots and trimmings, and put them in a cool oven to dry thoroughly; then grate the root, and rub the leaves and stalks through a sieve, and put all into a tightly corked bottle, or tin can with close cover; this makes a most delicious seasoning for soups, stews, and stuffing. When you use parsley, save every bit of leaf, stalk or root you do not need, and treat them in the same way as the celery. Remember in using parsley that the root has even a stronger flavor than the leaves, and do not waste a bit.

Dried Herbs.—When you buy a bunch of dried herbs rub the leaves through a sieve, and bottle them tightly until you need them; tie the stalks together and save them until you want to make what the French call a *bouquet*, for a soup or stew. A *bouquet* of herbs is made [Pg 19] by tying together a few sprigs of parsley, thyme and two bay-leaves. The bay-leaves, which have the flavor of laurel, can be bought at any German grocery, or drug-store, enough to last for a long time for five cents.

Table Sauce.—There is no reason why you should not sometimes have a nice relish for cold meat when you can make a pint of it for six cents, so I will give you a receipt for it. Get at Washington market at the herb stand, a bunch of TARRAGON; it will cost five cents in the summer, when it is green and strong, and not much more in the winter; put it in an earthen bowl, and pour on it one pint of scalding hot vinegar; cover it and let it stand until the next day; then strain it, and put it into a bottle which you must cork tight. Either put more hot vinegar on the tarragon, or dry it, and save it until you want to make more; you can make a gallon of sauce from one bunch, only every time you use it you must let it stand a day longer.

Celery Salt.—If you mix celery root, which has been dried and grated as above, with one fourth its quantity of salt, it makes a nice seasoning and keeps a long time.

Spice Salt.—You can make this very nicely by drying, powdering and mixing by repeated siftings the following ingredients: one quarter of an ounce each of powdered thyme, bay-leaf, and pepper; one eighth of an ounce each of marjoram and cayenne pepper; one half of an ounce each of powdered clove and nutmeg; to every four ounces of this powder add one ounce of salt, and keep the mixture in an air-tight vessel. One ounce of it added to three pounds of stuffing, or forcemeat of any kind, makes a delicious seasoning.

Lemon and Orange Tincture.—Never throw away lemon or orange peel; cut the yellow outside off carefully, and put it into a tightly corked bottle with enough alcohol to cover it. Let it stand until the alcohol is a bright yellow, then pour it off, bottle it tight, and use it for flavoring when you make rice pudding. Add lemon and alcohol as often as you have it, and you will always have a nice flavoring.

Vanilla Tincture.—Make this from a broken Vanilla Bean, just as you would make Lemon Tincture. When you make a plain rice pudding, and when you boil rice with sweetening, put a teaspoonful of either of these tinctures with it, and it will be very good.

Measuring.—Be careful about measuring. Do not think you can guess just right every time; you cannot do it. One day the dinner [Pg 20] will be a great deal better than another, and you will wonder why; it will be because it is carefully seasoned and properly cooked. A good rule for seasoning soups and stews, is half an ounce, or a level tablespoonful of salt, and half a level teaspoonful of pepper to each quart of water; try it, if it is right you will know how much to use; if it is not right, alter it to suit your taste; but settle the point for once, and then you will know what to depend upon. The following table will give you some good hints about measuring; there are four teaspoonfuls in one tablespoon; two tablespoonfuls in one ounce; two ounces in one wineglassful; two wineglassfuls in one gill; two gills in one good sized cupful; two cupfuls in one pint; two pints in one quart. One quart of sifted flour, thrown into the measure, and shaken down, but not pressed, weighs one pound; one quart of Indian corn meal, shaken down in the measure weighs one pound and three ounces; one quart of fine sugar weighs one pound and a half.

CHAPTER III.

BEVERAGES.

In my little book on "FIFTEEN CENT DINNERS," I decidedly advocate the substitution of milk or milk and water as a drink at meal times, for tea and coffee, on the score of economy; because milk is a food, while the two former drinks are chiefly stimulants. They are pleasant because they warm and exhilarate, but they are luxuries because they give no strength; therefore their use is extravagant when we are pinched for healthy food. It is true that when we drink them we do not feel as hungry as we do without them. The sensation of hunger is nature's sure sign that the body needs a new supply of food because the last has been exhausted; the change of the nourishing qualities of food into strength is always going on as long as any remains in the system; the use of tea, coffee, and alcohol, hinders this change, and consequently we are less hungry when we use them than when we do without them. Tea and coffee are certainly important aids to the cheerfulness and comfort of home; and when the first stage of economy, where every penny must be counted, has passed, we do [Pg 21] not know of any pleasanter accessory to a meal than a cup of good tea or coffee.

Tea.—The physiological action of very strong tea is marked; moderately used it excites the action of the skin, lungs, and nervous system, and soothes any undue action of the heart; used to excess, it causes indigestion, nervousness, and wakefulness. No doubt its effects are greatly modified by climate, for the Russians drink enormous quantities of very strong, fine tea. A recent war report gives the following account of its use in the army.

"The Russian soldiers are said to live and fight almost wholly upon tea. The Cossacks often carry it about in the shape of bricks, or rather tiles, which, before hardening, are soaked in sheep's blood and boiled in milk, with the addition of flour, butter and salt, so as to constitute a kind of soup. The passion of the Russian for this beverage is simply astonishing. In the depth of winter he will empty twenty cups in succession, at nearly boiling point, until he perspires at every pore, and then, in a state of excitement rush out, roll in the snow, get up and go on to the next similar place of entertainment. So with the army. With every group or circle of tents travels the invariable tea kettle, suspended from a tripod; and it would be in vain to think of computing how many times each soldier's pannikin is filled upon a halt. It is his first idea. Frequently he carries it cold in a copper case as a solace upon the march."

Dr. Edward Smith sums up the physiological action of tea as follows:

1—A sense of wakefulness.

2—Clearness of mind, and activity of thought and imagination.

3—Increased disposition to make muscular exertion.

4 — Reaction, with a sense of exhaustion in the morning following the preceding efforts, and in proportion to them."

Coffee.— The action of coffee is so similar to that of tea that we need not consider it separately; it will be sufficient to remark that the chief points of difference are lessening the action of the skin, increasing the action of the heart, and, when used very strong, aiding digestion to some extent.

Cocoa and Chocolate.— Both these articles are made from the kernels of a tropical fruit, about the size of a cucumber, the fleshy part of which is sometimes used to produce a vinous liquor; they are produced [Pg 22] from the seeds of the cocoa palm, and from a kind of ground nut. These kernels consist of gum, starch, and vegetable oil; and are marketed as cocoa shells, which are the husks of the kernel; cocoa nibs, which consist of the crushed nuts; and ground cocoa, which is the kernels ground fine.

Chocolate is the finely ground powder from the kernels, mixed to a stiff paste with sugar, and, sometimes, a little starch. It is very nutritious; when it is difficult to digest remove from its solution the oily cake which will collect upon the surface as it cools. It is so nutritious that a small cake of it, weighing about two ounces, will satisfy hunger; for that reason it is a good lunch for travellers.

Both cocoa and chocolate are very nutritious, and are free from the reactionary influences of tea and coffee. Let us count the cost of these beverages, and see which is the best for us.

One quart of weak tea can be made from three teaspoonfuls, or half an ounce, of tea, (which cost at least one cent;) we must have for general use a gill of milk, (at one cent,) and four teaspoonfuls or one ounce of sugar, (at one cent); thus if we use only the above quantities of milk and sugar, one quart of tea costs three cents; if we increase them it will cost more.

One quart of weak coffee can be made from one ounce, or two tablespoonfuls of coffee, (at a cost of two cents;) two tablespoonfuls or ounces of sugar, (two cents,) and a half a pint of milk, (two cents;) the total cost six cents.

One quart of cocoa can be made from two ounces, or eight tablespoonfuls of cocoa shells, (which cost two cents,) with half a pint of milk, and an ounce of sugar, (at four cents more;) we have a quart of good, nutritious drink at six cents. It is all the better if the shells are boiled gently two or three hours. Of course the nibs, or crushed cocoa, and chocolate, will both produce a correspondingly nutritious beverage.

Beer.—Very poor families sometimes spend every day for beer enough to buy them a good, wholesome meal, because they think it makes them strong. Beer, like all other liquors, is of no value whatever in making strength; it only nerves you up to spend all you can muster under the excitement it causes, and then leaves you weaker than before. What you need when you crave liquor is a good, warm meal. The best doctors say that a man cannot drink more than about [Pg 23] a pint and a half of beer a day without injuring his health; and that healthy people, during youth and middle age, do not need it at all. Let it, and all other liquors alone entirely, and you will be better off in health and purse.

Beer for Nursing Women.—It is generally believed that women who drink malt liquor are able to nurse children to greater advantage than those who do not use it. The fact is that while the quantity of milk may be increased, its nourishing quality will be impaired. There may be more milk for the child, but it will be poor. The effect of all malt liquors is to promote the secretion of the fluids of the body, but not to enrich them. Do not drink beer for the sake of your child, but try milk, or milk and water instead, and see if after a fair trial you do not have plenty for the baby, and if it does not grow strong and fat. If milk does not agree with you, or you cannot afford it, use barley water; it will not only give you plenty of milk, but it will nourish you as well as the baby. You will get from it all the nourishment that you may fancy you get from malt liquor, with this advantage: in the barley water you will get all the nutriment of the grain unchanged, while in the form of beer the fermentation has destroyed part of it. The following is a good receipt:

Barley Water.—Thoroughly wash two ounces of pearl barley, (which costs less than two cents,) to remove any musty or bad flavor, put it over the fire in two quarts of cold water, and boil it until it is reduced to one quart; then strain it, cool it, and drink it whenever you are thirsty. A little sugar can be used without hurting the baby.

Milk.—I have already said that milk is the most perfect food; we will now see what it is made of, and how it nourishes the body; and then we can understand how necessary it is to have it pure. The elements of milk which strengthen the whole body are the solid parts that separate in the form of curd when it begins to turn sour; the whey contains the salts and phosphates which strengthen the brain, bones, and digestive organs; the cream is the part which makes us fat. When we remember that cheese is made from the curd of milk we can see why it is so valuable as food, and why a meal of black or brown bread and cheese will satisfy a hungry man.

Try to afford at least a quart of good milk every day. It can be bought in New York now for seven or eight cents a quart; and if the [Pg 24] children have plenty of seconds bread, or oatmeal porridge, and a cup of milk, at meal times, they will be strong and rosy.

Skim-milk, butter-milk, and whey, are all excellent foods, and far better drinks than beer or whiskey. Make a plain pudding now and then, with skim-milk, adding an ounce of suet to restore its richness. If the milk has turned a little sour add lime water to it, in the proportion of four tablespoonfuls of the lime water to a quart. If the lime water is added before the milk begins to turn it will help keep it fresh. The following is a good receipt for making lime water:

Lime Water.—Slack four ounces of quick lime with a little water, and gradually add enough water to make a gallon in all; let it stand three hours, then bottle it in glass-stoppered bottles, putting a portion of the undissolved lime in each bottle; when you want to use some, pour off the clear fluid from the top.

Children should never have tea, coffee, or liquor; all these drinks hurt them; give them milk, or milk and water; or pure water, if you cannot afford milk. But you had better scant their clothes than their supply of milk. If you have to limit the supply of food, deny them something else, but give them plenty of bread and scalded milk, and you can keep them healthy.

CHAPTER IV.
BREAD, MACARONI, AND RICE.

Homemade bread is healthier, satisfies hunger better, and is cheaper than bakers' bread. Make bread yourself if you possibly can. Use "middlings" if you can possibly get them; they contain the best elements of wheat. "Household Flour" has similar qualities, but is sometimes made from inferior kinds of wheat. Both are darker and cheaper than fine white flour; and bread made from them takes longer to "rise" than that made from fine flour. Bakers' bread is generally made from poor flour mixed with a little of the better sort; or with a little alum, which added to the wheat grown in wet seasons, keeps the bread from being pasty and poor in taste.

The prices of bakers' bread upon the streets in the eastern and [Pg 25] western parts of the city are as follows: ordinary white bread, five cent loaf weighs three quarters of a pound: six cent loaf weighs fourteen ounces: eight cent loaf weighs one pound and ten ounces; black bread, two eight cent loaves weigh, respectively, one pound eight, and one pound ten ounces; fine French bread, eight cent loaf weighs three quarters of a pound; in the French quarter a six cent loaf weighs one pound. We advise the purchase of new flour in preference to old, because, unless flour is cooled and dried before it is packed, the combined action of heat and dampness destroys its gluten, and turns it sour; gluten is the nutritive part of the flour, that which makes it absorb water, and yield more bread. If you do not have a good oven, your bread can be baked at the baker's for about a cent a loaf. When bread is made too light it is tasteless, and lacks nourishment, because the decay caused in the elements of the flour used to make it by the great quantity of yeast employed, destroys the most nutritious parts of it. A pint of milk in a batch of four loaves of bread gives you a pound more bread of better quality, and helps to make it moist. Scalded skim milk will go as far as fresh whole milk, and you can use the cream for some other dish. One pound of pea-meal, or ground split-peas, added to every fourteen pounds of flour used for bread increases its nourishment, and helps to satisfy hunger.

Keep your bread in a covered earthen jar; when it is too stale to eat, or make into bread broth, dry it in a cool oven, or over the top of the fire, roll it with a rolling-pin, sift it through a sieve, and save the finest crumbs to roll fish or chops in for frying, and the largest for puddings. If a whole loaf is stale put it into a tight tin can, and either steam it, or put it into a moderately warm oven for half an hour; it will then be as good as fresh bread to the taste, and a great deal more healthy.

A good allowance of bread each day is as follows: for a man two pounds, costing six cents; for boys and women one pound and a half, costing five cents; for children a pound each, costing three cents.

Homemade Bread.—Put seven pounds of flour into a deep pan, and make a hollow in the centre; into this put one quart of luke-warm water, one tablespoonful of salt, one teaspoonful of sugar, and half a gill of yeast; have ready three pints more of warm water, and use as much of it as is necessary to make a rather soft dough, mixing and kneading it well with both hands. When it is smooth and shining [Pg 26] strew a little flour upon it, lay a large towel over it folded, and set it in a warm place by the fire for four or five hours to rise; then knead it again for fifteen minutes, cover it with the towel, and set it to rise once more; then divide it into two or four loaves, and bake it in a quick oven. This quantity of flour will make eight pounds of bread, and will require one hour's baking to two pounds of dough. It will cost about thirty cents, and will last about two days and a half for a family of six. In cold weather, the dough should be mixed in a warm room, and not allowed to cool while rising; if it does not rise well, set the pan containing it over a large vessel of boiling water; it is best to mix the bread at night, and let it rise till morning, in a warm and even temperature.

Rice Bread.—Simmer one pound of rice in three quarts of water until the rice is soft, and the water evaporated or absorbed; let it cool until it is only luke-warm; mix into it nearly four pounds of flour, two teaspoonfuls of salt, and four tablespoonfuls of yeast; knead it until it is smooth and shining, let it rise before the fire, make it up into loaves with the little flour reserved from the four pounds, and bake it thoroughly. It will cost about twenty-five cents, and make more than eight pounds of excellent bread.

Potato Bread.—Take good, mealy boiled potatoes, in the proportion of one-third of the quantity of flour you propose to use, pass them through a coarse sieve into the flour, using a wooden spoon and adding enough cold water to enable you to pass them through readily; use the proper quantity of yeast, salt, and water, and make up the bread in the usual way. It will cost about twenty-four cents if you use the above quantities, and give you eight pounds or more of good bread.

Pulled Bread.—Take from the oven an ordinary loaf of bread when it is about *half baked*, and with the fingers, *while it is yet hot*, pull it apart in egg-sized pieces of irregular shape; throw them upon tins, and bake them in a slow oven to a rich brown color. This bread is excellent to eat with cheese or wine. An ordinary sized loaf, costing about three cents makes a large panful.

Bread made with Baking Powder.—Where bread is made with baking powder the following rules should be closely observed: if any shortening be used, it should be rubbed into the flour before it is wet; *cold* water or sweet milk should always be used to wet it, and the dough should be kneaded immediately, and only long enough to thoroughly [Pg 27] mix it and form it in the desired shape; it should then be placed in a well-heated oven and baked quickly—otherwise the carbonic acid gas will escape before the expanded cells are fixed in the bread, and thus the lightness of the loaf will be impaired.

Breakfast Rolls.—Mix well by sifting together half a pound of flour, (cost two cents,) a heaping teaspoonful of baking powder, a level teaspoonful of salt, and a heaping teaspoonful of sugar, (cost one cent;) rub into a little of the above one ounce of lard, (cost one cent,) mix it with the rest of the flour, and quickly wet it up with enough cold milk to enable you to roll it out about half an inch thick, (cost two cents;) cut out the dough with a tin shape or with a sharp knife, in the form of diamonds, lightly wet the top with water, and double them half over. Put them upon a tin, buttered and warmed, and bake them in a hot oven. This receipt will cost about six cents, and will make about nine good sized rolls.

Tea Biscuit.—Mix as above, using the same proportions, and cutting out with a biscuit-cutter; when they are baked, wash them over with cold milk, and return them to the oven for a moment to dry. The cost is the same.

Macaroni.—This is a paste made from the purest wheat flour and water; it is generally known as a rather luxurious dish among the wealthy; but it should become one of the chief foods of the people, for it contains more gluten, or the nutritious portion of wheat, than bread. It is one of the most wholesome and economical of foods, and can be varied so as to give a succession of palatable dishes at a very small cost. The imported macaroni can be bought at Italian stores for about fifteen cents a pound; and that quantity when boiled yields nearly four times its bulk, if it has been manufactured for any length of time. Good macaroni is yellow or brownish in color; white sorts are always poor. It should never be soaked or washed before boiling, or put into cold or lukewarm water; wipe it carefully, break it in whatever lengths you want it, and put it into boiling water, to every quart of which half a tablespoonful of salt is added; you can boil an onion with it if you like the flavor; as soon as it is tender enough to yield easily when pressed between the fingers, drain it in a colander, saving its liquor for the next day's broth, and lay it in cold water until you want to use it. When more macaroni has been boiled than is used it can be kept perfectly good by laying it in fresh water, which must be changed every day. [Pg 28] After boiling the macaroni as above, you can use it according to any of the following directions. Half a pound of uncooked macaroni will make a large dishful.

Macaroni, Farmers' Style.—Boil half a pound of macaroni as above, and while you are draining it from the cold water, stir together over the fire one ounce each of butter and flour, and as soon as they bubble gradually pour into the sauce they make, a pint of boiling water, beating it with a fork or egg whip until it is smooth; season it with a level teaspoonful of salt and a level saltspoonful of pepper, and put the macaroni in it to heat; then cut an onion in small shreds, and brown it over the fire in a very little fat; when both are done dish the macaroni, and pour the onion out of the frying pan upon it. It is excellent; and ten cents will cover the cost of all of it.

Macaroni with Broth.—Put half a pound of macaroni, boiled as above and washed in cold water, over the fire with any kind of broth, or one pint of cold gravy and water; season it to taste with pepper and salt, and let it heat slowly for an hour, or less if you are in a hurry; then lay it on a flat dish, strew over it a few bread crumbs, which you will almost always have on hand if you save all the bits I speak of in the article on BREAD; then set the dish in the oven, or in front of the fire to brown. It will cost less than ten cents, and be delicious and very hearty.

Macaroni with White Sauce.—Warm half a pound of macaroni, boiled and washed in cold water, as above, in the following sauce, and use it as soon as it is hot. Stir together over the fire one ounce each of butter and flour, pouring in one pint of boiling water and milk, as soon as the butter and flour are mixed; season it with salt and pepper to taste, and put the macaroni into it. This dish costs less than ten cents, and is very good and wholesome.

Macaroni with Cheese.—Boil half a pound of macaroni, as above, put into a pudding dish in layers with quarter of a pound of cheese, (cost four cents,) grated and mixed between the layers; season it with pepper and salt to taste; put a very little butter and some bread crumbs over it, and brown it in the oven. It will make just as hearty and strengthening a meal as meat, and will cost about twelve cents.

Macaroni Milanaise Style.—Have ready some sauce, made according to the receipt for *Tomato Sauce* given below, or use some fresh tomatoes passed through a sieve with a wooden spoon, and highly sea [Pg 29] soned, and two ounces of grated cheese; put half a pound of imported Italian macaroni, (cost eight cents,) in three quarts of boiling water, with two tablespoonfuls of salt, one saltspoonful of pepper, and a teaspoonful of butter, (cost one cent;) boil rapidly for about twenty minutes, then drain it in a colander, run plenty of cold water from the faucet through it, and lay it in a pan of cold water until you are ready to use it. Put into a sauce-pan one gill of tomato sauce, (cost two cents,) one ounce of butter, (cost two cents,) and one gill of any meat gravy free from fat, and stir until they are smoothly blended: put a half inch layer of macaroni on the bottom of a dish, moisten it with four tablespoonfuls of the sauce, sprinkle over it half an ounce of the grated cheese; make three other layers like this, using all the macaroni, cheese, and sauce, and brown the macaroni in a hot oven for about five minutes; serve it hot. It will cost about thirteen cents.

Macaroni with Tomato Sauce.—Boil half a pound of macaroni as already directed, and lay it in cold water. Make a tomato sauce as follows, and dress the macaroni with it, using only enough to moisten it, and sprinkling the top with an ounce of grated cheese, (cost one cent;) serve it hot.

Tomato Sauce.—Boil together for one hour quarter of a can of tomatoes, or six large fresh ones, (cost five cents,) one gill of broth of any kind, one sprig of thyme, one sprig of parsley, three whole cloves, three peppercorns, and half an ounce of onion sliced; (cost two cents;) rub them through a sieve with a wooden spoon, and set the sauce to keep hot; mix together over the fire one ounce of butter, and half an ounce of flour, (cost two cents,) and when smooth incorporate with the tomato sauce. The cost of the tomato sauce will be about ten cents, and of the entire dish about eighteen cents. If you do not wish to use all the tomato sauce, and you do not need to, save it to use with fried chops of any kind.

Rice.—Rice is largely composed of starch, and for that reason is less nutritious than flour, oatmeal, Indian meal, or macaroni; but it is a wholesome and economical food when used with a little meat broth, drippings, or molasses. It is a very safe food for children, especially if used with a little molasses. The following is an excellent supper dish. [Pg 30]

Rice Panada.—Boil half a pound of rice, (which costs five cents,) quarter of a pound of suet, (at two cents,) with one tablespoonful of salt, and one of sugar, (cost one cent,) fast in boiling water for fifteen minutes; meantime mix half a pound of flour, (cost two cents,) gradually with one quart of water, and one gill of molasses, (cost two cents;) stir this into the boiling rice, and boil it for about five minutes; this makes a nice supper of over five pounds of good, nutritious food for twelve cents.

Boiled Rice.—Another good dish of rice for supper can be made as follows. Wash half a pound of rice (cost five cents,) throw it into one quart of boiling water, containing two teaspoonfuls of salt, and boil it fast ten minutes; drain it in a colander, saving the water to use with broth next day; meantime just grease the pot with sweet drippings, put the rice back in it, cover it, and set it on a brick on the top of the stove, or in a cool oven, and let it stand ten minutes to swell; be careful not to burn it. The addition of a very little butter, sugar, molasses, nutmeg, lemon juice, or salt and pepper, will give it different flavors; so that you can vary the taste, and have it often without getting tired of it, and it need never cost you over seven cents.

Rice, Milanaise Style.—Fry one ounce of butter, (cost two cents,) light brown; put into it half a pound of rice, (cost five cents,) well picked over, *but not washed*, and one ounce of onion, chopped fine; stir and brown for about five minutes, then add a pint of gravy from meat, season with a level teaspoonful of salt, quarter that quantity of pepper, and as much cayenne as you can take on the point of a very small pen-knife blade; the onion and seasoning will cost less than two cents; stew gently for fifteen minutes, stirring occasionally to prevent burning, and serve as soon as the rice is tender. This makes a palatable dish for about ten cents.

Rice, Japanese Style.—Put half a pound of well washed rice into a double kettle, with one pint of milk or water, one heaping teaspoonful of salt, and quarter of a medium sized nutmeg grated; boil it until tender, about forty minutes; if it seems very dry add a little more liquid, taking care not to have it sloppy when it is cooked. When milk is used it may be served with milk and sugar as a breakfast or tea dish; when water takes the place of milk, the addition of an ounce of butter, and half a saltspoonful of pepper makes a nice dinner dish of it. [Pg 31]

CHAPTER V.

SOUP.

The value of soup as food cannot be overestimated.

In times of scarcity and distress, when the question has arisen of how to feed the largest number of persons upon the least quantity of food, the aliment chosen has always been soup. There are two reasons for this: first, by the addition of water to the ingredients used we secure the aid of this important agent in distributing nutrition equally throughout the blood, to await final absorption; and, second, we gain that sense of repletion so necessary to the satisfaction of hunger—the fact being acknowledged that the sensation we call hunger is often allayed by the presence of even innutritious substances in the stomach.

Good soup is literally the juice of any ingredient from which it is made—the extract of the meat, grains, or vegetables composing it. The most economical of soups, eaten with bread, will satisfy the hunger of the hardest worker. The absolute nutritive value of soup depends, of course, upon its ingredients; and these can easily be chosen in reference to the maintenance of health. For instance, the pot-liquor in which meat has been boiled needs only the addition of a few dumplings or cereals, and seasoning, to form a perfect nutriment. That produced from skin and bones can be made equally palatable and nutritious by boiling with it a few vegetables and sweet herbs, and some rice, barley, or oatmeal. Even the gelatinous residue produced by long-continued boiling, without the presence of any foreign matter, is a useful emollient application to the inflamed mucous surfaces in some diseases, while it affords at the same time the degree of distention necessary to prevent flatulency.

The time required to make the most palatable and nutritious soup is short. Lean meat should be chopped fine, placed in cold water, in the proportion of a pint to each pound, slowly heated, and thoroughly skimmed. Five minutes' boiling will extract from the meat every particle of its nutriment and flavor. The liquor can then be strained off, [Pg 32] seasoned, and eaten with bread, biscuit, or vegetables. Peas or beans boiled and added to the soup make it the most perfect food for sustaining health and strength. It is the pure juice of the meat and contains all its savory and life-giving principles.

If your family is large, it will be well for you to keep a clean saucepan, or pot on the back of the stove to receive all the clean scraps of meat, bones, and remains of poultry and game, which are found in every kitchen; but vegetables should not be put into it, as they are apt to sour. The proper proportions for soup are one pound of meat and bone to one and a half quarts of cold water; the meat and bones to be well chopped and broken up, and put over the fire in cold water, being brought slowly to a boil, and carefully skimmed as often as any scum rises; and being maintained at a steady boiling point from two to six hours, as time permits; one hour before the stock is done, add to it one carrot and one turnip pared, one onion stuck with three cloves, and a bouquet of sweet herbs.

When soup is to be boiled six hours you must allow two quarts of water to every pound of meat, and you must see that the pot boils slowly and regularly, and is well skimmed. When you want to keep soup from one meal to another, or over night, you must pour it into an earthen pot, or bowl, because it will turn by being allowed to remain in the metal pot.

I shall give you first some receipts for making soups without meat, and then some of the cheapest meat soups I have tried. The first is very cheap and nutritious, and should be served at meals where no meat is to be used; bread, and a cheap pudding, will be sufficient to use with it.

Scotch Broth without Meat.—Steep four ounces of pearl barley, (cost three cents,) over night in cold water, and wash it well in fresh water; cut in dice half an inch square, six ounces of yellow turnip, six ounces of carrot, four ounces of onion, two ounces of celery, or use in its place quarter of a saltspoonful of celery seed, (cost of all about one cent,) put all these into two and a half quarts of boiling water, season with a teaspoonful of salt, quarter of a saltspoonful of pepper, and as much cayenne as you can take up on the point of a very small pen-knife blade, (cost one cent;) boil slowly for two hours; then stir in quarter of a pound of oatmeal, (cost two cents,) mixed to a smooth batter with cold water, see if seasoning be correct, add two or three [Pg 33] grates of nutmeg, and boil half an hour. Meantime, cut two slices of bread, (cost one cent,) in half inch dice, fry light brown in hot fat, (cost two cents,) and lay the bits in the soup tureen; when the soup is ready pour it over them, and serve. This soup, which costs only about ten cents, is palatable as well as economical.

Pea Soup.—Use half a pint, or seven ounces of dried peas, (cost three cents,) for every two quarts of soup you want. Put them in three quarts of cold water, after washing them well; bring them slowly to a boil; add a bone, or bit of ham, if you have it to spare, one turnip, and one carrot peeled, one onion stuck with three cloves, (cost three cents,) and simmer three hours, stirring occasionally to prevent burning; then pass the soup through a sieve with the aid of a potato-masher, and if it shows any sign of settling stir into it one tablespoonful each of butter and flour mixed together dry, (cost two cents;) this will prevent settling; meantime fry some dice of stale bread, about two slices, cut half an inch square, in hot fat, drain them on a sieve, and put them in the bottom of the soup tureen in which the pea soup is served; or cut some bits of very hard stale bread, or dry toast, to use instead of the fried bread. By the time the soup is done it will have boiled down to two quarts, and will be very thick and good. This receipt will cost you about ten cents.

Thick Pea Soup.—Fry one sliced onion, (cost half a cent,) in one ounce of suet or drippings, (cost half a cent,) using an iron pot to fry it in; as soon as it is brown, put into the same pot, three quarts of cold water, one pint, or fourteen ounces of well washed peas, (cost five cents,) and boil as above; this quantity of peas does not need any crusts in the soup; it will be thick enough; but bread may be eaten with it, if you want it. This soup costs six cents.

Bean Soup.—For this, use the receipt for pea soup, using beans instead of peas; the cost will be about the same.

Lentil Soup.—For two quarts of soup half a pint of yellow lentils, (cost five cents,) washed, and put to boil in three pints of cold water, with one cents' worth of soup greens, and boiled gently until the lentils are soft enough to break between the fingers; every half hour a gill of cold water should be added, and the lentils again raised to a boiling point, until they are done; they should then be passed through a sieve with a wooden spoon, using enough of the liquor to make them pass easy, and mixed with the rest of the soup; it should be seasoned with [Pg 34] salt and pepper, and is then ready to simmer for half an hour, and serve hot, with dice of fried bread half an inch square, like those used for pea soup, or with bits of stale bread. A plentiful dinner of lentil soup and bread costs only about ten cents.

Onion Soup.—Chop half a quart of onions, (cost three cents,) fry them brown, in a large saucepan, with two ounces of drippings, stirring until they are well browned, but not burned; then stir in half a pound, or a little less, of oatmeal, (cost three cents,) add three quarts of water, and season to taste with pepper and salt; (the drippings and seasoning cost one cent;) while the soup is boiling, which must be for about twenty minutes, with occasional stirring, toast a third of a six cent loaf of bread, cut it in half inch bits, lay it in the soup tureen; and, when the soup is ready pour it on the toast. The soup will cost about ten cents, and is extremely nourishing.

Spinach Soup.—Put one quart of spinach, (cost five cents,) to boil in a large pot, full of boiling water, well salted with two tablespoonfuls of salt; cover until it boils up once; then remove the cover, and with a wooden spoon press the spinach under water as fast as it rises to the surface; boil it steadily only until it is tender; then drain it; run plenty of cold water from the faucet over it, while it is still in the colander; drain it again, chop it fine, and pass it through a kitchen sieve with the aid of a wooden spoon; boil one quart of milk, (cost eight cents,) and one quart of water; add the spinach to it, thicken it by stirring in two tablespoonfuls of corn starch dissolved in cold milk; season it with one teaspoonful of salt, quarter of a saltspoonful of white pepper, and the same of nutmeg; (cost of seasoning one cent,) and serve it as soon as it boils up. It costs only fifteen cents, and is delicious.

Soup can be made from any green vegetable or herb in the same way.

Vegetable Soup.—The following is the receipt given by the celebrated FRANCATELLI for a cheap vegetable soup: Put six quarts of water to boil in a large pot with quarter of a pound of suet, or two ounces of drippings, (cost about two cents,) season it with a level tablespoonful of salt, half a teaspoonful of pepper, and a few sprigs of parsley and dried herbs, (cost of seasoning one cent;) while it is boiling prepare about ten cents' worth of cabbage, turnips, beans, or any cheap vegetables in season; throw them into the boiling soup, and when they [Pg 35] have boiled up thoroughly, set the pot at the side of the fire, where it will simmer, for about two hours. Then take up some of the vegetables without breaking, and use them with any gravy you may have on hand, or with quarter of a pound of bacon, (cost four cents,) sliced and fried, for the bulk of the meal; the soup after being seasoned to taste can be eaten with bread, at the beginning of the meal, the whole of which can be provided for about twenty cents.

Vegetable Porridge.—Pare and slice thin ten cents' worth of carrots, turnips, onions, and parsnips, and put them into three quarts of water, with a few sprigs of parsley and dried herbs; season them with half a tablespoonful of salt, and quarter of a teaspoonful of pepper, and let them boil till very soft, two hours or more; then rub them all through a colander, return the porridge to the pot, and set it over the fire to heat, stirring it to prevent burning. Use it with bread; it will cost about fifteen cents for enough for a hearty meal.

Rice Milk.—Put half a pound of well-washed rice into two quarts of boiling water, with two ounces of sweet drippings, a teaspoonful of salt, and a bit of cinnamon, or lemon peel, and let all boil gently about an hour; then add one quart of milk, and stir the rice for about ten minutes. A little sugar or molasses may be added if you want it sweet. It makes an excellent breakfast or supper dish, and costs about fifteen cents.

Fish Soup.—Make this soup from any rich, glutinous fish, such as cod's head, halibut neck, flounders, skate, or any cheap fish which is in season, and which you can buy for five or six cents a pound. Chop one or two onions, fry them in a pot with two ounces of drippings, till light brown; season with a level tablespoonful of salt, half a teaspoonful of pepper, and a teaspoonful of sweet herbs of any kind, then add two quarts of hot water, and let all boil for ten minutes; meantime mix quarter of a pound of oatmeal with one pint of cold water, and wash and cut in two-inch pieces about two pounds of fish; when the soup has boiled ten minutes, put the fish into it, and carefully stir in the oatmeal; let it boil twenty minutes, stirring occasionally to prevent burning; it will then be ready to use. The seasoning, drippings, and oatmeal, will cost about five cents, and the fish ten more; with the addition of bread and potatoes, say five cents' worth of either, it makes an excellent meal, costing about twenty cents.

Fish Chowder.—Fry together in the bottom of a saucepan four [Pg 36] ounces of salt pork and two onions sliced; when brown season with a teaspoonful of sweet herbs, and a very little salt and pepper; meantime peel and slice half a dozen medium-sized potatoes, and lay them in cold water; and cut in small slices two pounds of any fish which costs about five cents per pound; when the onions and pork are brown, put the potatoes and fish upon them in layers, seasoning with a level tablespoonful of salt, and half a teaspoonful of pepper; pour over all cold water enough to cover the ingredients, and let them cook twenty minutes after they begin to boil; soak half a pound of sea-biscuit in cold water, and when the chowder is nearly done lay them on it, and pour over them half a pint of milk; in five minutes the chowder will be ready to use. The onions, pork, and seasoning will cost five cents; the potatoes, crackers and milk five more; and the fish ten cents; total for two quarts of good chowder twenty cents.

Mutton Broth.—Put two pounds of a jointed neck of mutton, (cost twelve cents,) in two and a half quarts of cold water, and let it boil slowly; skim it carefully, season it with a level tablespoonful of salt, half a teaspoonful of pepper, and the same of sweet herbs; then add one quart of yellow turnips, peeled and quartered, (cost three cents,) and four ounces of well washed pearl barley, (cost two cents,) and boil about an hour longer, or until the turnips and barley are tender. Take up the meat on a platter, lay the turnips around it, and pour the broth and barley into a soup tureen. The broth, meat and vegetables will cost seventeen cents, and will make a good dinner with the addition of bread; or you can use the mutton and turnips for one meal, and keep the broth and barley for another.

Veal Broth.—Make this as you would the mutton broth, using a knuckle of veal, (which costs ten cents,) instead of the neck of mutton, and a quarter of a pound of rice, (which costs two or three cents,) instead of barley; omit the turnips. You will have a good, nutritious, broth for about thirteen cents.

White Broth.—Cut two pounds of the neck of veal, (cost twelve cents,) in cutlets, and put it in a sauce pan with two ounces of salt pork, (cost two cents,) a level tablespoonful of salt, half a teaspoonful of pepper, one onion chopped, six whole cloves, and half a pint of water; (the seasoning will cost about one cent;) boil these ingredients for ten minutes, stirring often enough to prevent burning, then add two and a half quarts of hot water, and skim the broth thoroughly as soon [Pg 37] as it boils up; let it simmer for half an hour, when take up the meat, reserving it for stew, strain the broth, let it boil up again, and then put into it a quarter of a pound of macaroni, (cost four cents,) and boil it for half an hour longer. While it is boiling put the meat with half a quart of peeled and quartered potatoes, (cost two cents,) a teaspoonful of salt, and a pint of boiling water into a sauce pan and let them cook as long as the macaroni. Serve the stew by itself, and the broth and macaroni in a soup tureen. With bread these two dishes make a good dinner, at a cost of about twenty-five cents. You can sometimes use rice or dumplings instead of macaroni.

Cream Soup.—Proceed as for white broth, using the meat for a stew, skimming all the fat off the broth, and then adding to it two tablespoonfuls of flour mixed smooth with half a pint of milk; when the milk and flour are mixed smooth pour into them a gill of the boiling broth, and then add them to the soup; see if the seasoning is right, and boil it ten minutes, stirring it to prevent burning; during this time toast a few slices of stale bread, cut them in dice, and put them in the soup tureen; when the soup is ready pour it over the toast, take up the stew on another dish, and serve both together. They make a good dinner for about twenty-five cents.

Beef Broth.—Put two pounds of lean beef from the neck, (cost twelve cents,) in two and a half quarts of cold water to boil; skim as soon as it boils, and add a level tablespoonful of salt, half a teaspoonful of pepper, quarter of a nutmeg grated, a few sweet herbs, and half a dozen cloves; (cost of seasoning two cents;) boil gently for one hour. At the end of quarter of an hour make as follows some

Norfolk Dumplings.—Mix by sifting together one pound of flour, one teaspoonful of salt, and two of baking-powder, (cost three cents;) make into a soft dough with one egg, half a pint of milk and a very little water, (cost four cents,) and drop them by the tablespoonful in the soup; be careful that the pot does not stop boiling, or the dumplings will be heavy.

At the end of three quarters of an hour stir together over the fire in a large sauce-pan one ounce each of butter and flour, (cost two cents,) and when they are nicely browned, gradually add, and mix with an egg-whip or large fork, a pint of the boiling soup. Take up the meat and dumplings on the same dish, strain the soup into the sauce you have just made, and mix it thoroughly; put a little of it over [Pg 38] the meat and dumplings, and serve the rest in the soup tureen; it is very nice with small dice of toast in it.

Both dishes make an excellent dinner, at a cost of about twenty-five cents, including bread.

Meat Brewis.—This palatable dish is made by putting the thick upper or under crust of a stale loaf of bread into the pot when any meat or soup is boiling, and letting it remain about five minutes to thoroughly absorb the broth; it should then be taken out as whole as possible, and eaten at once. It is satisfactory, nutritious and economical; enough for a hearty meal costing not more than five cents.

CHAPTER VI.

PEAS, BEANS, LENTILS, AND MAIZE.

Before giving you receipts for cooking peas, beans, and lentils, I want to show you how important they are as foods. I have already spoken of the heat and flesh forming properties of food as the test of its usefulness; try to understand that a laboring man needs twelve ounces and a half of heat food, and half an ounce of flesh-food every day to keep him healthy. One pound, or one and a quarter pints of dried peas, beans, or lentils, contains nearly six ounces of heat food, and half an ounce of flesh food; that is, nearly as much heat-food, and more than twice as much flesh food as wheat. A little fat, salt meat, or suet, cooked with them, to bring up their amount of heat-food to the right point, makes either of them the best and most strengthening food a workingman can have. The only objection to their frequent use is the fact that their skins are sometimes hard to digest; but if you make them into soup, or pudding, rubbing them through a sieve after they are partly cooked, you will be safe from any danger.

Oatmeal and Peas.—Cut quarter of a pound of fat pork or bacon, (cost four cents,) into pieces half an inch square; put it in the bottom of a pot with two sliced onions, (cost one cent,) and fry ten minutes without burning; season with two teaspoonfuls of salt, one of sugar, and one saltspoonful of pepper; (cost of seasoning one cent;) then add three quarts of cold water, and one pint of peas, (cost five cents,) and [Pg 39] boil the whole gently until the peas become quite soft; then stir in enough oatmeal to thicken, about a quarter of a pound, (cost two cents or less;) simmer for twenty minutes, and then eat hot. It is the healthiest kind of a meal, and costs thirteen cents, or less.

Peas-Pudding.—Soak one pint of dried peas, (cost five cents,) in cold water over night; tie them loosely in a clean cloth, and boil them about two hours in pot-liquor or water, putting them into it cold and bringing them gradually to a boil; drain them, pass them through a sieve with a wooden spoon, season them with a level tablespoonful of salt, half a saltspoonful of pepper, one ounce of butter, and one egg, (all of which will cost five cents,) mix, tie in a clean cloth, and boil half an hour longer; then turn it from the cloth on a dish, and serve hot. This receipt makes a good large pudding for ten cents; or you can leave out the egg and it will cost less.

Peas and Bacon.—Put one pound of bacon, (cost twelve cents,) to boil in two and a half quarts of cold water, with one pint of dried peas, (cost five cents;) when the peas are soft, drain them, press them through a sieve, lay them neatly on a flat dish, place the bacon on them, and set them in the oven to brown. Meantime strain any water which may remain after boiling them, and thicken it over the fire with Indian meal, in the proportion of four or five tablespoonfuls to each pint, so as to make it thick enough to cut and fry when cold; boil it about one hour, and then cool it.

As soon as the peas and bacon are brown, serve them with boiled potatoes or bread, (about five cents' worth of either;) they make a good dinner, and with the hasty pudding, cost only about twenty-five cents.

Baked Peas.—Proceed as directed for peas-pudding, only instead of putting the peas again in the cloth put them in a pudding-dish, and brown them in the oven. A large dish costs only ten cents.

Peas and Onions.—Proceed as directed for peas pudding, omitting the egg, and substituting for it an onion chopped and fried in an ounce of drippings; bake as in the previous receipt. The cost will be about ten cents, and the dish is exceedingly nutritious.

Baked Beans.—Put one pint of dried beans, (cost six cents,) and quarter of a pound of salt pork, (cost four cents,) into two quarts of cold water; bring them to a boil, and boil them slowly for about twenty minutes; then put the beans, with about a teacupful of the water they were boiled in, into an open jar, season them with salt and [Pg 40] pepper to taste, and one tablespoonful of molasses, (cost of seasoning one cent,) lay the pork on the top, and bake two hours, or longer. The dish will cost about ten cents, and is palatable and nutritious. The liquor in which the beans were boiled should be saved, and used the next morning as broth, with seasoning and a little fried or toasted bread in it.

Stewed Beans.—Soak a pint of dried beans, (cost six cents,) over night in cold water; put them to boil in a quart of cold water with one ounce of drippings, a level tablespoonful of salt, and quarter of a teaspoonful of pepper, and boil them gently for two hours. Then drain them, put them into a sauce pan with one ounce of butter and a tablespoonful of chopped parsley, and after heating them through, serve them at once. The drippings, butter, and seasoning, will not cost more than four cents, and the whole dish can be made for ten.

Fried Beans.—Proceed as above, omitting the parsley, and letting the butter get hot in a frying pan, before putting the beans in; fry them a little, stirring them so that they will brown equally, and then serve them. The dish will cost ten cents.

Beans and Bacon.—Soak a pint of dried beans, (cost six cents,) over night in cold water; put them over the fire in one quart of cold water, with quarter of a pound of bacon, (cost three cents,) and boil them gently for about two hours; then stir in two tablespoonfuls of flour mixed smooth with a little cold water, season to taste with pepper, salt, and if you like it a little chopped onion, and let them stew gently for about ten minutes; they will then be ready to serve. The dish will cost ten cents.

Lentils.—Lentils have been used for food in older countries for a long time, and it is quite necessary that we should become acquainted with their merits if we want to save; I give a lentil soup, and some excellent directions for cooking this invaluable food. One quart of lentils when cooked will make four pounds of hearty food. There are two varieties in market; the small flat brown seed, called lentils *à la reine*; and a larger kind, about the size of peas, and of a greenish color; both sorts are equally well flavored and nutritious; they cost ten cents a pound, and can be bought at general groceries. The seed of the lentil tare, commonly cultivated in France and Germany as an article of food, ranks nearly as high as meat as a valuable food, being capable of sustaining life and vigor for a long time; this vegetable is [Pg 41] gradually becoming known in this country, from the use of it by our French and German citizens; and from its nutritive value it deserves to rank as high as our favorite New England Beans.

Lentils boiled plain.—Wash one pound, or one full pint of lentils, (cost ten cents,) well in cold water, put them over the fire, in three quarts of cold water with one ounce of drippings, one tablespoonful of salt, and a saltspoonful of pepper, (cost about one cent,) and boil slowly until tender, that is about three hours; drain off the little water which remains, add to the lentils one ounce of butter, a tablespoonful of chopped parsley, a teaspoonful of sugar, and a little more salt and pepper if required, (cost about three cents,) and serve them hot. Always save the water in which they are boiled; with the addition of a little thickening and seasoning, it makes a very nourishing soup.

Stewed Lentils.—Put a pint of plain boiled lentils into a sauce pan, cover them with any kind of pot-liquor, add one ounce of chopped onion, two ounces of drippings, quarter of an ounce of chopped parsley, and stew gently for twenty minutes; serve hot. This dish costs about ten cents.

Fried Lentils.—Fry one ounce of chopped onion brown in two ounces of drippings, add one pint of plain boiled lentils, see if they are properly seasoned, and brown them well; serve hot. This dish costs about ten cents, and is very good, and as nutritious as meat.

Maize, or Indian Corn Meal.—This native product is a strong and nutritious food, and very economical; in addition to the ordinary hasty-pudding, or mush, it can be cooked with a little pot-liquor, meat, or cheese, so as to be both good and wholesome. Below are some excellent receipts for cooking it.

Polenta.—Boil one pound of yellow Indian meal, (cost four cents,) for half an hour, in two quarts of pot-liquor or boiling water, salted to taste, with one ounce of fat, stirring it occasionally to prevent burning; then bake it for half an hour in a greased baking dish, and serve it either hot, or, when cold, slice it and fry it in smoking hot fat. This favorite Italian dish is closely allied to the hasty-pudding of New England, and the mush of the South. It costs five cents.

Cheese Pudding.—Into two quarts of boiling water, containing two tablespoonfuls of salt, stir one pound of yellow Indian meal, (cost four cents,) and a quarter of a pound of grated cheese, (cost four cents;) boil it for twenty minutes, stirring it occasionally to prevent burning; [Pg 42] then put it in a greased baking pan, sprinkle over the top quarter of a pound of grated cheese, (cost four cents,) and brown in a quick oven. Serve hot. If any remains, slice it cold and fry it brown. It costs twelve cents.

Hasty-Pudding.—Have boiling upon the fire two quarts of water with a level tablespoonful of salt; sprinkle in gradually one pound of Indian meal, (cost four cents,) stirring constantly to prevent lumps; and boil steadily for one hour, stirring occasionally. The secret of making good hasty-pudding is to boil it long enough to thoroughly cook it. Some persons first mix the meal with cold water until it forms a thick batter, and then stir this into the boiling water. The pudding can be eaten with a little milk, butter, or molasses, if they are desirable additions; or with some meat gravy, or melted and seasoned suet. When cold it is good sliced and fried.

Johnny Cake.—Mix one pound of Indian meal, (cost four cents,) one ounce of lard, (cost one cent,) and one teaspoonful of salt, with sufficient boiling water to make a stiff batter; put it by the tablespoonful into a greased baking pan, and bake it thoroughly. Five cents' worth makes a hearty meal, with a little butter or molasses.

Indian Cakes.—These are prepared in the same way as Johnny Cake, except that the batter is made about as thin as buckwheat cakes, and baked upon a greased griddle over the fire instead of in the oven. The most economical way of greasing the griddle is to put a small piece of fat salt pork upon a fork and rub it over the surface of the griddle after it is well heated.

Indian Bread.—Mix into one quart of boiling water enough Indian meal to make a thin batter, about a quarter of a pound, (cost one cent;) when it has cooled, stir into it one pound of wheat flour, (cost four cents,) a level tablespoonful of salt, and one gill of yeast; let it rise overnight, and then bake it in loaves.

Boiled Indian Pudding.—Dissolve a level teaspoonful of soda in one pint of sour milk, add to it one pint of molasses, (cost five cents,) quarter of a pound of chopped suet, (cost two cents,) half a pound of Indian meal, (cost two cents,) and a level teaspoonful of salt; if you have no milk use boiling water instead of it; put the pudding into a scalded pudding bag, or a pudding kettle, and this into a pot of boiling water; boil steadily for four hours, adding boiling water as the quantity [Pg 43] decreases. The pudding when cooked may be eaten with sauce or molasses, if desired; it will cost about ten cents.

Baked Indian Pudding.—Stir into a quart of boiling milk, (cost eight cents,) quarter of a pound of Indian meal, (cost one cent,) one level teaspoonful of salt, the same of spice, and one ounce of butter, (cost two cents;) last of all add one pint of cold milk, (cost four cents,) or milk and water. Pour the pudding into an earthen dish, and bake slowly for three hours. It will cost about fifteen cents, and be very nice.

There is as much difference in the quality of Indian meal as there is in its preparation; Southern meal is undoubtedly finer than Northern, and Southern cooks are proverbial for their skill in using it. I am indebted for some of the preceding receipts to a friend in Maryland, and I advise my readers to buy Southern meal, if they can get it, and test them thoroughly. Meal that is ground by hand or water power is superior to that ground by steam, because it is less heated in the process.

Indian corn is an excellent food in temperate and warm climates; and from its abundant yield, and easy cultivation, it is one of the cheapest of cereals. It contains the nitrates, or flesh-forming properties, in an excessive degree. It is a palatable and nutritious diet whether eaten green, parched, or ground into meal.

CHAPTER VII.
CHEAP FISH AND MEAT DINNERS.

I have already spoken of the value of fish as strengthening food, and in support of what I say I need only to remind you how vigorous and healthy the inhabitants of the sea coast usually are, especially if they eat red-blooded fish. This fact, in connection with the abundance and cheapness of fish makes it an important article in the dietary of the good housekeeper.

Fish may be cooked by boiling, baking, broiling, and frying; boiling is the least economical method of cooking fish, and fish soup, or [Pg 44] fish chowder the most saving; broiled fish wastes but little of its nutriment, but its bulk is decreased; baked fish ranks next to fish soup in point of economy.

Fish are preserved for winter use by cleaning them, rubbing them with salt, packing them in layers, and covering them with brine. An excellent way of pickling fish is to clean them, cut off the heads, tails, and fins, wash them, and then rub them well with salt and spice, pack them in layers in an earthen crock or deep dish, cover them with vinegar, and tie the jar over with buttered paper; they are then ready to bake slowly for about four hours; and will keep for three or four weeks after they are cooked.

In London, and other large English cities, where fried fish forms an important item of popular food, it is cooked with great care, and in such a manner as to retain all its nourishing qualities. It is well washed in salted water, dried on a clean cloth, cut in slices if large, dipped in a rather thin batter, made of flour, salt, pepper, and cold water, and then dropped into a pan containing plenty of fat heated until it is smoking hot, but does not boil; the pan is then taken from the fire, and by the time the fat is growing cool the fish is cooked. A novice would do best by maintaining the fat at the proper degree of heat until the fish is cooked.

The receipts which I give for fish are calculated to produce compound dishes from it, hearty enough to make the bulk of a meal.

Fish and Potato Pie.—Use any cheap fish which does not cost more than five or six cents a pound, such as cod, haddock, or bluefish; cut two pounds of fish, (cost twelve cents,) in pieces about an inch thick and two inches long; lay them in a deep dish with a pint of cold gravy of any kind, or cold water, season with a tablespoonful each of chopped parsley and onion, and a teaspoonful of salt, pepper, and thyme, mixed together in equal quantities, and sprinkled among the fish; put it into the oven for fifteen or twenty minutes to partly cook. Put one quart of potatoes, (cost three cents,) into boiling water, and boil until soft enough to mash; mash them, season them with salt and pepper, and put them over the fish, which you must take from the oven, as a crust; return the pie again to the oven to brown the crust, and then serve it with bread and butter. Twenty-five cents will cover the cost of all, and the dinner will be a good one.

Fish Pudding.—Make a plain paste by mixing quarter of a [Pg 45] pound of lard or sweet drippings, (cost three cents,) with half a pound of flour, (cost two cents,) a teaspoonful of salt, and just water enough to make a stiff paste; roll it out; line the edges of a deep pudding dish with it half way down; fill the dish with layers of fresh codfish cut in small pieces, using two pounds, (cost twelve cents,) season each layer with salt, pepper, chopped parsley, and chopped onions, using one tablespoonful of salt, one saltspoonful of pepper, two bay leaves, a saltspoonful of thyme, four ounces of onion, and half an ounce of parsley, (cost five cents;) fill up the dish with any cold gravy, milk, or water, cover with paste, and bake fifteen minutes in a quick oven; finish by baking half an hour in a moderate oven; serve hot.

With bread the dinner will cost twenty-five cents.

Fish and Potato Pudding.—Wash and peel one quart of potatoes, (cost three cents,) peel and slice about six ounces of onions, (cost one cent,) skin and bone two bloaters or large herrings, (cost five cents,) put all these ingredients into a baking dish in layers, seasoning them with a dessertspoonful of salt and a saltspoonful of pepper; pour over them any cold gravy you have on hand, or add two or three ounces of drippings; if you have neither of these, water will answer; bake the pudding an hour and a half; serve hot, with bread. If you use drippings or milk the entire seasoning will cost you less than five cents; and the whole dinner, which is excellent, not more than fifteen cents.

Codfish Steaks.—Two pounds of codfish, (which costs at the market from four to seven cents,) can be cut in steaks, dried well, and either dipped in flour, or thin batter of flour, salt, pepper, and water, and fried in smoking hot fat, or can be served with a quart of boiled potatoes, (cost three cents,) and plenty of bread and butter, at the rate of about twenty cents a meal.

Red Herrings with Potatoes.—Soak a dozen herrings, (cost ten cents,) in cold water for one hour; dry and skin them, split them down the back, and lay them in a pan with two ounces of drippings, two ounces of onion chopped fine, a saltspoonful of pepper, and three tablespoonfuls of vinegar, (cost two cents,) and set them in a moderate oven to brown for ten or fifteen minutes; meantime, boil one quart of potatoes, (cost three cents,) with a ring of the paring taken off, in plenty of boiling water and salt, pouring off the water as soon as they are tender, and letting them stand on the back of the fire, covered with a dry towel, for five minutes; serve them with the herrings, taking care to [Pg 46] dish both quite hot. With bread and butter a plentiful dinner can be had for about twenty-two cents.

Cheap Meats.—Those parts of meat which are called the cheap cuts, such as the head, brains, tongue, tripe, kidneys, haslet or pluck, feet, and tail, are eaten much more frequently in Europe than in this country, and are worthy of all the use they get there; for their proportion of flesh-forming elements is large; this is especially the case with the lights or lungs, but as they are rather difficult to digest, they should be thoroughly cooked, and never eaten alone. Tripe and pigs' feet, on the contrary, are very easily digested; but on this account are not as satisfactory food as that which remains longer in the stomach; although they are both savory and cheap.

Be careful to keep all meat stews closely covered, or a great deal of the nutriment of the meat will escape in the steam.

Sheeps' Head Stew.—Thoroughly clean a sheeps' head, weighing about three pounds, (cost about ten cents,) put it over the fire with quarter of a pound of rice, (cost three cents,) two cents' worth of onions sliced, a level tablespoonful of salt, quarter of a teaspoonful of pepper, and three pints of cold water; bring it slowly to a boil, skimming it carefully, and then add five cents' worth of carrots and turnips, peeled and quartered; let all simmer gently together for two hours, being careful to remove all grease, and see if the seasoning is correct, before dishing the stew. With bread, or boiled potatoes, the meal will cost about twenty-five cents.

Oxtail Stew.—Put two jointed oxtails, (cost about ten cents,) over the fire in one quart of cold water, and scald them, to remove the strong flavor; then roll the joints in flour, season them with salt and pepper, and pack them in an earthen jar, with one onion chopped, and one quart of potatoes peeled and sliced; the vegetables and seasoning will cost about five cents; add one pint of water, put on the cover of the jar, and cement it in place with a paste of flour and water, which you must grease a little to prevent cracking; then put the jar into a moderately hot oven, and bake it about four hours. With the addition of bread and butter it makes a hearty meal, and costs about twenty-two cents.

Beef Pie.—Cut in two inch pieces two pounds of the neck of beef, (cost twelve cents,) brown them quickly in one ounce of drippings, (cost one cent,) season them with pepper and salt, put them into a pud [Pg 47] ding dish in layers with one cents' worth of chopped onion, and one quart of potatoes, (cost three cents,) peeled and sliced; add enough cold water to cover the beef and vegetables, and put over them a crust made of one pound of flour, (cost four cents,) and quarter of a pound of lard, (cost three cents,) put it for fifteen minutes into a hot oven, and then bake for an hour and a half in a moderate one. It will cost less than twenty-five cents, and be an abundant meal.

Baked Heart.—Thoroughly wash a beef's heart, (cost ten cents,) stuff it with half a loaf of stale bread, (cost two cents,) moistened with warm water and seasoned with one teaspoonful of salt, quarter of a teaspoonful each of pepper, chopped parsley and sweet herbs, an onion chopped, and one ounce of sweet drippings (cost of all these two cents;) lay it in a dripping pan with five cents' worth of parsnips scraped and washed, and bake in a moderate oven about two hours. It may be baked in an earthen jar, like the oxtail stew, and all its goodness will be saved.

Parsnips are exceedingly nutritious and cheap, but if they are not liked potatoes may be substituted for them.

The entire dinner with bread and butter will cost about twenty-five cents.

Stewed Kidneys and Potatoes.—Wash one quart of potatoes, (cost three cents,) pare off one ring from each, and put them to boil in well salted boiling water. Choose a very fresh beef's kidney, (cost fifteen cents,) cut it in thin slices, removing all the white vessels and membranes, fry it quickly for five minutes in one ounce of smoking hot drippings, (cost one cent,) season it with half a teaspoonful of salt, and quarter of a teaspoonful of pepper, a teaspoonful each of chopped parsley, onion, and vinegar; shake into it from the dredging box one tablespoonful of flour, add one pint of boiling water, and boil gently for fifteen minutes. By this time the potatoes will be done, and both dishes must be served at once, because the kidneys will grow tough and indigestible if they are cooked more than twenty minutes in all. They will make a plentiful dinner, including bread and butter, for about twenty-five cents.

Pig's Kidneys may be cooked in the same manner, and enough can be bought for ten cents to make a good sized dish.

Kidney Pudding.—Cut the kidneys, season, and stew them as above, making meantime a crust from one pound of flour, two teaspoonfulls of [Pg 48] salt, and one of baking powder, sifted together; mix into these ingredients four ounces of finely chopped suet, (cost two cents,) make them into a paste with about one pint of cold water; use part of this to line a deep pudding dish, into which put the stewed kidneys; cover the dish with the rest of the paste, and bake it about an hour and a quarter in a regular, moderately hot oven. The pudding will cost about thirty cents.

Gammon Dumpling.—Make a plain paste of one pound of flour, (cost four cents,) one dessertspoonful of salt, and one of baking powder, quarter of a pound of finely chopped suet or scraps, (cost two cents,) and sufficient cold water to mix it to a stiff dough; roll this out about half an inch thick, spread over it half a pound of any cheap cut of bacon or ham, finely chopped, (cost six cents,) roll up the dumpling as you would a roly-poly pudding, tie it tightly in a clean cloth, and boil it in boiling water, or boiling pot-liquor, for about three hours. Do not let the pot stop boiling, or the dumpling will be heavy. Serve it hot, with one quart of plain boiled potatoes, (cost three cents.) The dinner will cost fifteen cents.

Bacon and Apple Roly-poly.—Boil a pound of bacon, (cost twelve cents,) for half an hour; then slice it thin; peel and slice three cents' worth of apples and the same quantity of onions; make a stiff dough of one pound of flour, (cost four cents,) a teaspoonful of salt, and cold water; roll it out half an inch thick; lay the bacon, apples, and onion all over it, roll it up, tie it tightly in a clean cloth, and boil it about two hours, in plenty of boiling water. Serve it with three cents' worth of boiled potatoes, or boiled cabbage. The dinner will cost twenty-five cents.

Mutton and Onions.—Choose a shoulder of mutton weighing about three pounds, which you can buy at six cents a pound; wipe it thoroughly with a damp cloth, put it into a pot half full of boiling water, with a tablespoonful of salt, and boil it gently for two hours, skimming the pot as often as any scum rises. Half an hour before it is done slice one quart of onions, (cost five cents,) boil them in a pint of boiling water for about twenty minutes, add one ounce of butter, (cost two cents,) half a pint of milk, (cost two cents,) four tablespoonfuls of flour (cost one cent,) one teaspoonful of salt, and pepper to taste. When you have put the onions over the fire, pare rings off a quart of potatoes, (cost three cents,) and boil them in well salted boiling water. Have [Pg 49] all three dishes ready at once, and serve them together hot. Save the broth from the mutton, and the next morning boil it up once, and serve it for breakfast, with half a loaf of stale bread, toasted, and cut in dice; or boil in it for twenty minutes a quarter of a pound of rice or macaroni.

The dinner will cost you about thirty cents, and you have on hand the broth for breakfast.

Pork and Onions.—Three pounds of the neck, or spare ribs, of fresh pork, which you can buy at the packing houses for three cents a pound, can be made into a capital dinner, which will cost only about twenty cents, by following the above receipt.

Veal and Rice.—Put the scrag end of a neck of veal, which you can usually buy for ten cents, into a pot half full of boiling water, with a half tablespoonful of salt, and half a pound of bacon, or salt pork, (cost six cents,) half a pound of rice, (cost five cents,) and an onion stuck with six cloves; boil it gently for three hours, and then serve it hot, the meat in the middle of the platter, and the rice laid around it. The broth may be served for breakfast, as in the receipt for MUTTON AND ONIONS.

The dinner will cost about twenty cents.

Irish Stew.—Cut two pounds of the flank of beef, (cost fifteen cents, or less,) in pieces about two inches square, rub them well with pepper and salt; peel and slice one quart of onions, (cost five cents;) place beef and onions in a saucepan, with just enough cold water to cover them, and stew them gently for one and a half hours; then add one quart of peeled potatoes, (cost three cents,) and boil the stew until the potatoes are soft, which will be in about twenty minutes. Serve at once hot. The dish will cost twenty-three cents.

Sheep's Haslet.—Peel and slice one quart of onions, (cost five cents;) wash and slice a sheep's haslet, (cost six cents;) put two ounces of drippings, (cost two cents,) in the bottom of a dripping pan, strew the onions upon it, and lay the haslet on them, seasoning it with a teaspoonful of salt, and one of thyme, savory, allspice, and pepper, using equal parts of each; add enough water to reach half-way to the top of the meat, then cover it thickly with the crumbs from half a loaf of stale bread, and bake all together for one hour and a half, in a moderate oven. The whole dish will not cost over seventeen cents, and it is nutritious and savory. [Pg 50]

Baked Pig's Head.—Buy at a packing house half a medium sized pig's head, which you can get for three or four cents a pound, (the piece will cost about ten cents;) clean and wash it well; pare and slice one quart of onions, (cost five cents;) chop quarter of a pound of suet, (cost two cents,) and grate half a loaf of stale bread, (cost three cents;) put into a dripping pan one ounce of drippings, (cost one cent,) one gill of vinegar, (cost one cent,) then the onions, next the head, skin up, and last the bread, suet, and seasoning, well mixed, and bake in a moderate oven for about one and a half hours. The dish will cost about twenty-two cents; it is hearty and extremely nutritious.

CHAPTER VIII.

SUNDAY DINNERS.

Sunday is the workingman's festival. It is not only a day of rest from manual labor, a breathing space in his struggle for existence, an interval during which his devotional aspirations may have full exercise; it is the forerunner of a new phase of life, in which toil is laid aside for the gentler occupations of home, if he is a man of family, and for rest and relaxation in any case.

The duty of making home pleasant, which a good wife feels, is doubly felt upon the days when the bread-winner abides in it. The husband of such a wife seldom passes his Sundays in strange places: he is content to accept the day according to its recognized signification, and when it has passed he is all the more ready to begin his daily work again. Because much of the comfort of home depends upon good and economical meals, and because Sunday dinners ought to be better than those of working days, we must make Monday dinners supplementary to them; the cost of Saturday night's marketing must be divided between the two days, in order to keep within our financial margin. Good examples of this management may be found in the receipts given in this chapter for ROAST FOWL and FRIED CHICKEN, À LA MODE BEEF and MEAT PATTIES, BOILED MUTTON and KROMESKYS, and ROAST VEAL and VEAL AND HAM PATTIES. These receipts show how by the exercise of a little judgment in buying, and economy in [Pg 51] managing food, we can have our Sunday fowl, or joint of meat, without incurring any expense unwarranted by the figures to which this little book confines us.

Roast Fowl.—You can generally buy a fowl for about a shilling a pound; it need not be tender, but it ought to be fleshy in order to furnish the basis for two meals. Choose a fowl which will cost fifty cents or less; pluck all the pin feathers, singe off the hairs with a piece of burning paper, or a little alcohol poured on a plate and lighted with a match; then wipe the fowl with a clean damp cloth, draw it carefully by slitting the skin at the back of the neck, and taking out the crop without tearing the skin of the breast; loosen the heart, liver, and lungs by introducing the fore-finger at the neck, and then draw them, with the entrails, from the vent. Unless you have broken the gall, or the entrails, in drawing the bird, *do not wash it*, for this greatly impairs the flavor, and partly destroys the nourishing qualities of the flesh. Twist the tips of the wings back under the shoulders; bend the legs as far up toward the breast as possible, secure the thigh bones in that position by a trussing cord or skewer; then bring the legs down, and fasten them close to the vent. Put the bird into a pot containing three quarts of boiling water, with one tablespoonful of salt, an onion stuck with half a dozen cloves, and a bouquet of sweet herbs, made as directed on page 19; skim it as soon as it boils, and as often as any scum rises. If you wish to stuff the fowl use a forcemeat made as follows, (cost ten cents,) and carefully sew it up in the carcass.

Forcemeat or Stuffing.—Cut two ounces of salt pork, (cost two cents,) in quarter inch dice, and fry it brown in half an ounce of drippings, with one ounce of chopped onion; while these ingredients are frying, soak five cents' worth of stale bread in tepid water, and then wring it dry in a napkin; add it to the onion when it is brown, with one tablespoonful of chopped parsley, half a saltspoonful of powdered thyme, and the same quantity of dried and powdered celery, and white pepper, and one teaspoonful of salt; mix all these over the fire until they are scalding hot, and cleave from the pan; then stir in one raw egg, and stuff the fowl with it. It is good stuffing for any kind of poultry or meat. A few ounces of grated cheese make it superlatively good.

Meantime, while the fowl is boiling, peel one quart of potatoes, (cost three cents,) and lay them in cold water. At the end of one [Pg 52] hour take the fowl from the pot, taking care to strain and save the pot liquor, put it into a dripping pan with the potatoes, season them both with a teaspoonful of salt, and quarter of a teaspoonful of pepper, and put them in a rather quick oven to bake for about one hour. When both are well done, and nicely browned, take them up on hot dishes, and keep them hot while you make the following gravy:

Chicken Gravy.—Pour one pint of boiling water into the dripping pan in which the fowl was baked; while it is boiling up mix one heaping tablespoonful, or one ounce, of flour with half a cup of cold water, and stir it smoothly into the gravy; season it to taste with pepper and salt, and send it in a bowl to the table with the chicken and potatoes.

In carving the chicken cut off the drumsticks, wings, and neck carefully, and lay them aside; use the second joints, breast and fleshy parts, for dinner; and after dinner cut up what remains of the carcass in neat pieces, which you must save with the pieces first cut off, to use for FRIED CHICKEN.

Half the cost of the Roast Chicken, stuffed, and the Baked Potatoes, will be thirty-eight cents.

Fried Chicken.—Dip the pieces of chicken saved from the Sunday dinner into a batter made according to the following receipt, and fry it a delicate brown color in quarter of a pound of olive oil or sweet drippings, or lard, (cost three cents,) heated until it is smoking hot. Before you begin to fry the chicken, wash one quart of potatoes, (cost three cents,) pare off a ring from each, and put them to boil in plenty of well salted boiling water. When the chicken is done take it up with a strainer, and lay it for a few minutes on brown paper to free it from fat; then serve it hot, with the boiled potatoes.

Frying Batter.—This batter will do nicely for chicken, fish, clams, cold boiled parsnips, or fruit of any kind, of which you wish to make fritters. The oil is added to it for the purpose of making it crisp. Many persons object to the use of oil in cooking, from a most foolish prejudice. It is a pure vegetable fat, wholesome and nutritious in the highest degree; and the sooner our American housewives learn to use it in cooking the better it will be for both health and purse. I do not mean the expensive oil, sold at fine grocery stores for a dollar a bottle, but a good sweet kind which can be bought at French *Épicerie* or German *Delicatessen* depots for about two dollars and fifty cents a gallon. Make the batter by mixing together four heaping tablespoonfuls of [Pg 53] flour, (cost one cent,) a level teaspoonful of salt, the yolk of one egg, (cost one or two cents,) two tablespoonfuls of oil, (cost one cent,) and one gill of water, or a quantity sufficient to make a thick batter; just as you are ready to use it, beat the white of the egg, and stir it into the batter; the cost will be three or four cents, and the use of it will double the size and nicety of your dish.

Chicken Broth.—Heat the broth in which the fowl for Sunday dinner was boiled, and when it is at the boiling point throw in quarter of a pound of rice, or fine macaroni, which will cost three or four cents, and boil it about twenty minutes, or until tender; see if the seasoning is right, and serve it hot.

New York Cooking School Fricassee.—Prepare a fowl weighing about three pounds, (cost three shillings,) as directed in the receipt for Roast Fowl; cut it in neat joints, fry it quickly in one ounce of sweet drippings, (cost one cent,) till brown; cover it with boiling water, add one teaspoonful of salt, and quarter of a level teaspoonful of pepper, and stew it gently until tender, keeping it covered closely; when it is about half done, add to it some dumplings made as follows:

Suet Dumplings.—Make into a stiff paste, with about two gills of cold water, half a pound of flour, (cost two cents,) quarter of a pound of chopped suet, (cost two cents,) a teaspoonful of salt, and the same quantity of baking powder sifted with the flour; drop the paste into the fricassee from a teaspoon dipped in cold water, and let them boil with it; these dumplings cost less than five cents, and are nice with any stew, soup, or fricassee.

Rabbit Curry.—Choose a tender rabbit or hare, which will cost at the market about twenty cents, and which if young will be plump, and have a short neck, thick knees, and fore paws whose joints break easily; hang it by the hind legs, and skin it, beginning at the tail, and ending at the head, wipe it carefully with a damp cloth to remove the hairs; take out the entrails, saving the brains, heart and liver, rinse out the carcass with a cup of vinegar, (cost two cents,) which you must save, and cut it in joints; lay the rabbit in a deep frying pan, with two ounces of drippings, (cost two cents,) one cent's worth of onion sliced, a teaspoonful of salt, ten whole cloves, and quarter of a level teaspoonful of pepper; fry it gently for twenty minutes; then add one cent's worth of parsley, the vinegar, half a level tablespoonful of curry, and one tablespoonful of flour mixed with half a teacupful of water, and [Pg 54] simmer all gently for fifteen minutes, keeping the pan closely covered. When the rabbit is first put upon the fire, put quarter of a pound of rice, (cost four cents,) into two quarts of boiling water with one tablespoonful of salt, and boil it until the ends of the grains begin to crack open; turn it from the pot into a colander, drain it, shake it back into the pot, and cover it to keep it hot until the rabbit is done; then send it to the table with the rabbit, but on a dish by itself. The RABBIT CURRY AND RICE will cost about twenty-eight cents.

Rabbit Pie.—Prepare a rabbit, or hare, (cost twenty cents,) as for the CURRY, and after you have jointed it, roll each piece in flour, salt and pepper mixed; slice two cent's worth of onions, peel and slice three cents' worth of potatoes, and put these into a pudding dish in layers with the rabbit, season with a teaspoonful of salt, and quarter of a level teaspoonful of pepper, add half a pint of cold water, cover the pie with a plain paste, made as for SUET DUMPLINGS (cost five cents,) and bake for one hour and a quarter. These quantities will cost about thirty cents, and make a large pie.

Pickled Shad.—In season fine large shad can be bought for twenty-five cents, and each one will be enough for two hearty meals. Thoroughly clean a fresh shad; cut it in pieces about three inches square, lay it in a deep baking dish, or earthen crock, seasoning it well with two tablespoonfuls of salt, one level teaspoonful of pepper, one dozen whole cloves, two bay-leaves broken, and a bit of lemon or orange peel, if you have it; pour over it enough vinegar to cover it, tie an oiled or buttered paper over the top of the dish or crock, and bake the shad five hours in a moderate oven. The action of the pickle will be to entirely soften the bones of the fish, so that every part of it will be eatable. One half of it will cost about fifteen cents; and with the addition of five cents' worth of bread or potatoes, will make a hearty dinner for twenty cents.

Pork Pie.—Cut in two inch pieces two pounds of pork trimmings, (cost ten cents,) roll them in flour, season them with two teaspoonfuls of salt, quarter of a level teaspoonful of pepper, and one teaspoonful of curry, put them in a deep baking pan or dish with two cents' worth of onions, and three cents' worth of potatoes, peeled and sliced, add half a pint of cold water, and bake the pie slowly for one hour and a quarter. It will cost about fifteen cents made as above; or a suet crust, made as directed for SUET DUMPLINGS, may be added for five cents more. If the taste of curry is not liked it may be omitted. [Pg 55]

Pork Chops.—Buy at a packing house two pounds of shoulder chops, (cost sixteen to twenty cents,) roll them in flour, pepper, and salt, put them into a hot frying pan, and fry them brown, cooking them at least twenty minutes. Meantime boil one quart of potatoes, (cost three cents,) in boiling water and salt, and chop fine one pickle, (cost one cent.) When the chops are done, take them up, and keep them hot, while you make the gravy by pouring into the frying-pan half a pint of boiling water, and adding to it the chopped pickle, a tablespoonful of flour mixed smooth with half a cup of cold water, and salt and pepper to taste. Boil it up once, pour it over the chops, and serve them hot with the potatoes.

The dinner will cost about twenty-five cents.

Roast Pork and Apples.—Season two pounds of shoulder chops, (cost twenty cents, or less,) with salt and pepper, and powdered sage, and put them in a deep baking dish with one quart of potatoes, (cost three cents,) two cents' worth of onions, and two cents' worth of apples, peeled and sliced; add half a pint of cold water, and bake two hours in a moderate oven.

The dish will cost twenty-seven cents, or less.

Stewed Sausage.—Prick a pound and a half of sausages, (cost eighteen cents,) lay them in hot water for three minutes, roll them in flour, put them in a hot frying pan, and fry them brown; take them up and fry about half a loaf of stale bread sliced, in the same pan; put this on a platter, lay the sausages on it, and pour over them a gravy made as follows; after taking up the sausages, pour into the pan half a pint of boiling water, season it to taste with salt and pepper, thicken it with one tablespoonful of flour mixed smooth in half a cupful of cold water, add to it one chopped pickle, boil it up, and pour it over the sausages and bread. The seasoning and flour will cost two cents, the bread three, and the whole dish about twenty-three cents. If you serve it with a quart of plain boiled potatoes it will cost twenty-five or twenty-six cents.

German Potatoes.—Carefully wash one quart of potatoes, removing any defective part, cut a slice from the top of the potatoes, take out a little of the inside, chop it fine, mix it with half a pound of highly seasoned sausage or mincemeat, (cost six cents,) fill it into the potatoes, put on the piece you first cut off, and bake them for about three quarters of an hour in a quick oven. Serve them as soon as they are [Pg 56] soft. Ten cents will cover the entire cost, and they will make a very hearty and nutritious meal, especially if the meat used is pork.

Brain and Liver Pudding.—You can generally buy a pig's brain and haslet at the slaughter house for about ten cents; wash them thoroughly; slice the heart, liver, and lights, and fry them light brown in a cents' worth of drippings. Put the brain over the fire in cold water with a tablespoonful each of salt and vinegar, let it boil for fifteen minutes, and then lay it in cold water to get hard. Make a suet crust, as directed for SUET DUMPLINGS, (cost five cents,) roll out a cover for the pudding, line the edges of the dish two inches down with it, and put any bits you may have remaining, into the dish in layers with the haslet and brain sliced; season the pudding with one level tablespoonful of salt, one onion chopped, and half a level teaspoonful of pepper; cover it with the suet crust, and bake it for about an hour in a moderate oven. Serve it hot. The pudding will make a very hearty dinner, at a cost of about fifteen cents.

Broiled Kidneys.—Mix together in a deep plate the following ingredients, which will cost about three cents; one ounce of butter, half a level teaspoonful of pepper, one teaspoonful each of mustard, and any table sauce or vinegar, and as much cayenne as you can take up on the point of a small pen-knife blade; toast half a loaf of stale bread, (cost three cents,) cut in slices one inch thick; wash, split, and broil one pound of pig's or sheep's kidneys, (cost ten cents or less;) while the kidneys are broiling dip the toast in the first named seasonings, lay it on a hot dish, and lay the kidneys on it as soon as they are broiled; season them with salt and pepper, and serve them hot with one quart of plain boiled potatoes, (cost three cents.) The cost of the entire dinner will be less than twenty cents.

Tripe, Curry and Rice.—Thoroughly wash two pounds of tripe, (cost sixteen cents,) boil it until tender, about one hour, in plenty of water and salt; then lay it on a clean, dry cloth to drain; put half a pound of rice, (cost five cents,) into the same water, and boil it fast for twenty minutes; cut the tripe in pieces two inches square; slice two cents' worth of onions, frying them in two ounces of drippings, (cost two cents,) season with one teaspoonful of salt, quarter of a level teaspoonful of pepper, and one tablespoonful of vinegar, add to the tripe, and cook all together for fifteen minutes, stirring occasionally to prevent burning. Just as you are ready to serve it, stir in one teaspoon [Pg 57] ful of curry, which, with the other seasonings, will cost two cents. Drain the rice in a colander, shake it into a dish, and send it to the table with the tripe. The dinner will cost twenty-seven cents, and be very satisfactory.

Liver Polenta.—Boil one pound of yellow Indian Meal, (cost four cents,) for half an hour, in two quarts of boiling water with one ounce of drippings, (cost one cent,) stirring it occasionally to prevent burning; meantime fry in one ounce of drippings, (cost one cent,) a sheep's or pig's haslet, (cost five cents,) well washed and sliced; when the meal has boiled half an hour, put it into a greased baking dish with the haslet, seasoning each layer with salt and pepper; bake it for twenty minutes in a quick oven, and serve it hot.

The dish, which is palatable and nutritious, costs less than twelve cents.

À la Mode Beef.—This is one of the compound dishes which are mentioned in the beginning of this chapter, and will serve as a basis for at least two good dinners. Unless there is an unusual rise in the price of meat, you can buy the round of beef for a shilling a pound at the market or provision house; in the middle of the week choose four pounds in a solid, thick piece; cut half a pound of fat pork, (cost six cents,) into strips half an inch square; thrust the steel you use for sharpening knives into the meat, in the direction of the grain, and put the strips of pork into the holes you make; cut up five cents' worth of carrot, turnips, onion, and parsley, lay them in the bottom of an earthen crock or deep bowl, with two tablespoonfuls of salt, and one teaspoonful of pepper; put the beef on them, and pour over it one pint of vinegar, and enough water to just cover the meat; the vinegar and seasoning will cost five cents.

Turn a plate over the meat, and put a clean stone on it to keep the meat under the pickle; turn the meat every day, keeping it in a cool place.

Sunday morning, as soon as breakfast is over, put the meat, pickle, and vegetables, over the fire in a clean pot, and let them stew, *uncovered*, until the pickle is all evaporated and the meat is nicely browned; then sprinkle over it two tablespoonfuls of flour, and let that brown, turning the meat over occasionally; then add enough boiling water to cover the meat, put on the pot cover, and set it where it will simmer gently for at least three hours. During the last half hour boil one [Pg 58] quart of potatoes, (cost three cents,) in plenty of boiling water and salt. When the meat is done take it upon a platter, strain the gravy over it, and serve it hot with the boiled potatoes. About half of it will be enough for dinner, and will cost, with the potatoes, thirty-five cents.

Meat Patties.—Chop the remainder of the *À la mode* BEEF; make a suet crust, (cost five cents,) as directed for SUET DUMPLINGS, roll it out quarter of an inch thick, cut it out with a round tin cutter, lay a tablespoonful of the mince-meat on each round, wet the edges of the crust, and fold it over in the shape of an old-fashioned turnover; pinch the edges together, put the patties on a floured baking-pan, and bake them about half an hour in a moderate oven. When you put them in the oven, put one quart of potatoes, (cost three cents,) to boil in boiling water and salt. When both potatoes and patties are done serve them together; the dinner will cost about thirty cents.

Boiled Mutton.—The shoulder of mutton can be bought at the market for about six cents a pound. Choose one weighing not over four pounds, (cost twenty-four cents,) wipe it with a clean, damp cloth, put it into three quarts of boiling water with a tablespoonful of salt, one cents' worth of soup greens, a level teaspoonful of pepper, and boil it gently fifteen minutes for each pound, skimming it as often as any scum rises. About one hour before it is done pare one quart of turnips, cut them in quarters, and boil them with the mutton. Wash one quart of potatoes, pare off a ring from each, and boil them in boiling water. Serve them with the mutton and turnips, saving the broth from the mutton for BREAD BROTH for breakfast. The potatoes and turnips will cost five cents, and the proportionate cost of the mutton will be twelve cents; so the dinner will cost seventeen cents. The remains of the mutton must be saved for MUTTON *rechauffée*, as the basis of the next day's dinner.

Mutton *rechauffée*.—Prepare and boil one quart of potatoes, (cost three cents;) slice the best part of the mutton remaining from the day before, saving all the scraps and trimmings, dip each slice in a beaten egg, or a little milk, (cost one cent,) roll it in bread crumbs, dried and sifted, as directed on page 25, and fry them in sweet drippings. Serve the meat and potatoes together; they will cost about fifteen cents.

Mutton Kromeskys.—Cut cold mutton in half inch dice; chop one ounce of onion, and fry it pale yellow in one ounce of sweet drippings, (cost one cent;) add one ounce of flour, and stir until smooth; add half [Pg 59] a pint of water, two tablespoonfuls of chopped parsley, one level teaspoonful of salt, one level saltspoonful of white pepper, half a saltspoonful of powdered herbs, as much cayenne as can be taken up on the point of a very small penknife blade, and the chopped meat; the seasonings will cost about one cent; stir until scalding hot, add the yolk of one raw egg, (cost one cent,) cook for two minutes, stirring frequently; and turn out to cool on a flat dish, slightly oiled, or buttered, to prevent sticking, spreading the minced meat about an inch thick; set away to cool while the batter is being made.

Plain Frying Batter.—Mix quarter of a pound of flour, (cost one cent,) with the yolks of two raw eggs, (cost two cents,) a level saltspoonful of salt, half a saltspoonful of pepper, quarter of a saltspoonful of grated nutmeg, one tablespoonful of salad oil, (which is used to make the batter crisp,) and one cup of water, more or less, as the flour will take it up; the batter should be stiff enough to hold the drops from the spoon in shape when they are let fall upon it; now beat the whites of the two eggs to a stiff broth, beginning slowly, and increasing the speed until you are beating as fast as you can; the froth will surely come; then stir it lightly into the batter; heat the dish containing the meat a moment, to loosen it, and turn it out on the table, just dusted with powdered crackers; cut it in strips an inch wide and two inches long, roll them lightly under the palm of the hand, in the shape of corks, dip them in the batter, and fry them golden brown in smoking hot fat. Serve them on a neatly folded napkin. They make a delicious dish, really worth all the care taken in preparing them. The seasoning, crackers, and what fat is used in frying, will not cost over four cents, for you must strain the fat, and save it after you fry your KROMESKYS; if you use either bread or potatoes with them, the dinner will not cost over twenty cents.

Epigramme of Lamb.—This is one of my favorite dishes, which I learned to make the first winter I had a Cooking School, and I believe that nearly every one who tries it will share my opinion of it. Choose as tender a two-pound breast of mutton as you can buy for about six cents a pound, boil it in two quarts of water about three quarters of an hour, or until you can easily pull out the bones, taking care to put it into boiling water, with a tablespoonful of salt, and skim it as often as any scum rises; when it is done, strain and save the pot-liquor for BREAD or RICE BROTH, pull out the bones from the breast of mutton, [Pg 60] lay it between two platters, and put a flat iron on it until it is cold. Then cut it in triangular pieces, taking care not to waste a scrap, roll the pieces in a beaten egg, (cost one cent,) and dried bread crumbs prepared as directed on page 25, and fry them as you would the KROMESKYS in the previous receipt.

Use the pot-liquor in which it was boiled, with quarter of a pound of rice, for the next morning's breakfast. The cost of both dishes will not exceed twenty cents.

Roast Veal.—The shoulder of veal can usually be bought at the market for eight cents a pound. Choose a fresh one weighing about seven pounds, and costing about sixty cents; from this we shall make three dishes, namely: ROAST VEAL, BLANQUETTE OF VEAL, and VEAL AND HAM PATTIES. Therefore the proportionate cost for the ROAST VEAL will be twenty cents. Have the butcher chop off the fore leg quite close up to the shoulder, and cut it in neat slices about one inch thick; these you must sprinkle with salt and pepper, and keep in a cool place, together with the blade bone, until the next day, for the BLANQUETTE. Have the shoulder boned, saving the blade; stuff it with the following forcemeat.

Forcemeat for Veal or Poultry.—Steep four ounces of dry bread, (cost two cents,) in warm water, and wring it dry in a clean towel; chop one cent's worth of onion and fry it light yellow in one cent's worth of drippings, add the bread to it, season it with one level teaspoonful of salt, quarter of a level teaspoonful each of pepper and powdered thyme, or mixed spice, and stir these ingredients over the fire until they are scalding hot; then stir in one egg, and use the stuffing; the cost will be about five cents.

After stuffing the shoulder, lay it in a dripping pan with one cent's worth of soup greens, and put it in a hot oven to brown it quickly; when it is brown take it out of the oven, season with salt and pepper, baste it with a little sweet drippings, return it to the oven, and bake it thoroughly fifteen minutes to each pound. Meantime wash one quart of potatoes, (cost three cents,) pare a ring off each one, and boil them in plenty of boiling water and salt. When the veal is done take it up on a hot dish, pour half a pint of boiling water in the dripping pan, scrape it well, and strain the contents; set this gravy again over the fire to boil while you mix a tablespoonful of flour, in half a cup of cold [Pg 61] water; stir this smoothly into the gravy, boil it for five minutes, and serve it with the roast veal and boiled potatoes.

Be careful to save all that remains from the dinner, towards making the VEAL AND HAM PATTIES; the proportionate cost will be about thirty cents.

Blanquette Of Veal.—Put the pieces of veal saved for this dish into enough cold water to cover them, together with a tablespoonful of salt and one cent's worth of soup greens, the onion being stuck with ten cloves; skim occasionally whenever any scum rises, and simmer until the meat is tender, which will be in half or three quarters of an hour; then take up the meat in a colander, and run some cold water over it from the faucet; strain the pot-liquor, and let it boil again; mix together over the fire one tablespoonful of butter, (cost two cents,) and two of flour; when they are smooth add one quart of the boiling broth to them, season with a teaspoonful of salt, quarter of a level teaspoonful of white pepper, and quarter of a nutmeg grated; mix the yolks of two eggs, (cost two cents,) with about a cupful of the broth, and stir them into the rest; then put in the veal, and heat and serve it, with a quart of boiled potatoes, (cost three cents.) The dinner will cost about thirty cents.

Veal and Ham Patties.—Chop the remains of the ROAST VEAL (cost twenty cents,) with quarter of a pound of lean ham, (cost four cents,) weigh both, and mix with them an equal weight of dried bread, soaked in warm water, and wrung dry in a clean towel; season with salt, pepper, and powdered herbs, or SPICE SALT to taste, moisten with any cold gravy you have saved from the ROAST VEAL, and fill it into little turnovers, or patty pans lined with a suet crust, made as directed on page 53, for SUET DUMPLINGS, (cost five cents.)

The dinner will cost about thirty cents. [Pg 62]

CHAPTER IX.
CHEAP PUDDINGS, PIES, AND CAKES.

Good puddings are nutritious and wholesome, and an excellent variety can be made at a comparatively small expense. Pies, as they are usually made, with greasy and indigestible pastry, are positively unhealthy; if they are made with a plain bottom crust, and abundantly filled with ripe fresh or dried fruit, they are not so objectionable. Rich cake is always an extravagance, but some of the plainer kinds are pleasant additions to lunch and supper; we subjoin a few good receipts.

Swiss Pudding.—Sift together half a pound of flour, (cost two cents,) one heaping teaspoonful of baking powder, and one of salt; rub together four ounces of granulated sugar, (cost three cents,) and two ounces of butter, (cost four cents,) and when they are well mixed, so as to be granular but not creamy, add the flour gradually until all is used; make a hollow in the middle of the flour, put into it one egg, half a teaspoonful of lemon flavoring, and half a pint of milk, (cost of these ingredients four cents;) mix to a smooth paste, put into a well buttered and floured mould, and set this into a large pot with boiling water enough to come two-thirds up the side of the mould; steam the pudding three quarters of an hour, or until you can run a broom splint into it without finding the pudding stick to the splint. Turn the pudding out of the mould, and send it to the table with the following sauce:

Cream Sauce.—Stir together over the fire one ounce each of flour and butter, (cost two cents;) as soon as they are smooth pour into them half a pint of boiling milk, (cost two cents,) add two ounces of sugar and half a teaspoonful of lemon flavoring, (cost two cents,) and use with the pudding as soon as it boils up. The sauce and pudding will cost about twenty cents.

College Puddings.—Mix well together eight ounces of dried and sifted bread crumbs, (cost three cents,) two ounces of very finely chopped suet, (cost two cents,) two ounces of currants, two eggs, and two ounces of sugar, (cost together five cents,) a teaspoonful of salt, three grates [Pg 63] of nutmeg, and sufficient milk to moisten, about one cents' worth; make the puddings up in little round balls, roll them first in sifted bread crumbs; next dip them in beaten egg, and then roll them again in bread crumbs; fry them in plenty of hot fat, and serve them with sugar dusted over them. Five cents will cover the cost of frying them; and a nice dishful will cost you about eighteen cents.

Cream Rice Pudding.—Wash four ounces of rice, (cost three cents,) through two waters, put it into a baking dish with three ounces of sugar, and a teaspoonful of flavoring, (cost three cents,) pour in one quart and a pint of milk, (cost twelve cents,) and put it into a moderate oven to bake an hour and a half, or until it is of a creamy consistency. This pudding is very delicate and wholesome, and costs fifteen cents.

Half-pay Pudding.—Carefully wash and dry a quarter of a quarter of a pound of Zante currants, (cost three cents,) stone the same quantity of raisins, (cost three cents,) and chop an equal amount of suet, (cost two cents;) mix them with eight ounces of stale bread, (cost three cents,) three tablespoonfuls of molasses, half a pint of milk, and a teaspoonful each of spice, salt, and baking powder, (cost four cents.) Put these ingredients into a mould which has been well buttered and floured, and steam them about three hours. If by any mischance the top of the pudding is watery, you can remedy it by putting it into a hot oven for ten or fifteen minutes to brown. When you are ready to use it, turn it from the mould and send it to the table with some CREAM SAUCE. This is an excellent plum pudding, and costs only about twenty cents, including sauce.

Bread Pudding.—Slice a five cent loaf of bread, spread it lightly with two cents' worth of butter, and lay it in a baking dish; break one egg, (cost one cent,) into a bowl, add to it two ounces each of flour and sugar, (cost two cents,) a teaspoonful of salt, and a pint of milk, (cost four cents;) mix, flavor to taste, pour over the bread, and bake the pudding about half an hour in a quick oven. It will be very nice, and cost about fifteen cents.

Cup Custards.—Boil a pint of milk, (cost four cents,) with two ounces of sugar and half the yellow rind of a lemon, (cost three cents;) meantime beat four eggs, (cost four cents,) and strain the milk into them; mix thoroughly, strain again, and pour into cups; set these in a baking pan containing hot water enough to reach half way up the sides [Pg 64] of the cups, and either set the pan over the fire until the custards are firm, or bake them in the oven; they will set in twelve or fifteen minutes. The cost will be about twelve cents.

Fruit Dumpling.—Make a nice suet crust, as directed for SUET DUMPLINGS on page 53, roll it out about quarter of an inch thick, spread it with ten cents' worth of ripe fruit, quarter of a pound of sugar, (cost three cents,) and a teaspoonful of mixed spice; roll it up, tie it in a cloth wet with scalding water, and well floured next the dumpling, and boil it in a large kettle half full of boiling water for two hours, taking care that the pot does not stop boiling, or remain uncovered, or the dumpling will be heavy.

When it is done take it from the pot, unroll it from the cloth, and serve it with a few cents' worth of molasses; it will cost about twenty cents.

Apple Dumplings.—Pare and core five cents' worth of apples, keeping them whole; make a suet crust as directed for SUET DUMPLINGS on page 53, roll it out, and cut it in as many squares as you have apples; sprinkle a little spice on the apples, fold the corners of the pieces of paste up over them, pinch them together, tie each one in a floured cloth, and boil for one hour as directed in the previous receipt; then take them from the pudding cloths, and serve them with five cents' worth of butter and sugar; they can be made for about fifteen cents.

Baked Apple Dumplings.—Make a paste of half a pound of flour, (cost two cents,) quarter of a pound of butter, (cost eight cents,) and enough cold water to wet it up, about half a pint; roll it out very thin and fold it four times; repeat this process twice; then put the paste in a cool place for five minutes, and roll and fold again; do this three times, and then cut the paste in squares, and lay on each an apple prepared as above; fold the paste over the apples, turn them bottom up on a baking sheet, brush them with a well beaten egg, (cost one cent,) sift over them an ounce of powdered sugar, (cost one cent,) and put them in a moderate oven to bake for three quarters of an hour. They will cost about eighteen cents, and be very nice.

Lemon Dumplings.—Sift eight ounces of dried bread crumbs, (cost three cents,) mix them with the same quantity of very finely chopped suet, (cost four cents,) pare off the thin yellow rind of a lemon, (cost two cents,) chop it very fine, and add it with the juice to the bread and [Pg 65] suet; mix in half a pound of sugar, (cost six cents,) one egg, (cost one cent,) and enough milk to make a stiff paste, about half a pint, (cost two cents;) divide the paste into six equal balls, tie them in a floured cloth as directed for BOILED APPLE DUMPLINGS, and boil them an hour. Serve them with five cents' worth of butter and sugar, or syrup. They will cost about twenty-three cents, and are really delicious.

Rice Croquettes.—Boil half a pound of well washed rice, (cost five cents,) in one quart of cold water, with a level tablespoonful of salt, half a pint of milk, (cost two cents,) half the yellow rind of a lemon, or two inches of stick cinnamon, and two ounces of sugar, (cost two cents,) for half an hour, after it begins to boil, stirring it occasionally to prevent burning; take it from the fire, stir in one at a time, the yolks of three eggs, (cost three cents,) and return to the fire for ten minutes to set the egg; then spread the rice on an oiled platter, laying it about an inch thick, and let it get cool enough to handle. When it is cool enough turn it out of the platter upon some cracker dust spread on the table, cut it in strips one inch wide and three inches long, roll them into the shape of corks, dip them first in beaten egg, then in cracker-dust, and fry them golden brown in plenty of smoking hot fat; lay them on a napkin for a moment to free them from grease, put them on a dish, dust a little powdered sugar over them, and serve them. They will cost, including the last mentioned ingredients, about twenty cents.

Fruit Tarts.—Stew ten cents worth of fruit and four ounces of sugar together; make some pastry according to the directions in the receipt for BAKED APPLE DUMPLINGS; line deep pie-plates with the paste, building up a rim of paste around each; fill them with the stewed fruit, and bake them about three quarters of an hour in a moderate oven; two good sized tarts can be made for twenty-five cents; and the fruit can be varied to suit the season of the year, and the taste of the eaters.

Rice Cakes.—Sift together six ounces each of rice and wheat flour, (cost about seven cents,) rub into them four ounces of lard or meat drippings, (cost four cents,) four eggs, (cost four cents,) and sufficient milk to make a thick cake-batter; beat it thoroughly, pour it into a greased cake-pan, and bake it one hour. A good sized cake will cost about fifteen cents.

Rock Cakes.—Mix well together four ounces each of butter and sugar, (cost twelve cents,) add four ounces of well washed currants, (cost [Pg 66] three cents,) one pound of flour, (cost four cents,) and three eggs, (cost three cents;) beat all these ingredients thoroughly, roll them into little balls, or rocks, and bake them on a buttered baking pan. A good supply will cost about twenty-two cents.

Caraway Cake.—Beat to a cream four ounces each of butter and sugar, (cost twelve cents,) stir in two eggs, (cost two cents,) one gill of milk, (cost one cent,) one pound of sifted flour, (cost four cents,) and five cents' worth of caraway seed; bake the cake for two hours in a deep earthen dish, testing it with a clean broom splint to be sure it is done before you take it from the oven. It will cost about twenty-four cents.

Soft Gingerbread.—Melt one ounce of butter, (cost two cents,) add it to half a pint of molasses, (cost five cents,) with one level teaspoonful each of ground cloves, cinnamon, and ginger, (cost one cent;) dissolve one level teaspoonful of soda in half a pint of boiling water, mix this with the molasses, and lightly stir in half a pound of sifted flour (cost two cents;) line a cake-pan with buttered paper, pour in the batter, which will be very thin, and bake it about half an hour, or until you can run a broom-splint into it, and withdraw it clean. The cake, which will be a good size, will cost about ten cents.

Sweet Biscuits.—Rub four ounces of butter, (cost eight cents,) into one pound of flour, (cost four cents;) dissolve four ounces of sugar, (cost three cents,) in half a pint of warm milk, (cost two cents.) Pour this into the flour, mixing it smoothly; then dissolve half a level teaspoonful of cream of tartar in one gill of cold water, and stir it into the above ingredients. When they are thoroughly mixed, roll out the paste about quarter of an inch thick, cut it out in small round cakes, and bake them golden brown, at once, in a quick oven. A good supply will cost about seventeen cents. [Pg 67]

CHAPTER X.

DESSERT DISHES.

The previous chapter was devoted to cheap and good sweet dishes of the kind usually called dessert in this country; the dessert proper, however, consists of fruit, creams, ices, small and delicate cakes, fancy crackers, and confectionery. We give here directions for making some of these enjoyable delicacies at a very moderate rate.

It must always be borne in mind that the prices quoted are those which prevail when the articles specified are in season, and consequently abundant and cheap. As apples are very plentiful, and generally cheap, we shall begin with dishes made from them.

Apple Black Caps.—Pare a quart of nice apples, core them without breaking, set them side by side in a baking dish that will just hold them, fill the centres with sugar, place two cloves in the top of each one, grate over them the yellow rind of a lemon or orange, and put them into a moderate oven only until they are tender; do not let them break apart. As soon as they are tender take them from the oven, heat a fire shovel red hot and hold it over them, near enough to blacken their tops. Serve either hot or cold.

A porcelain-lined baking dish, or a *gratin* pan, is the best dish for cooking the black-caps in, because either can be set upon a clean plate and sent to the table; if the apples have to be removed from the dish in which they were baked they may be broken, and then the appearance of the dish will be spoiled.

The flavor of the dish may be changed by varying the spice, and by occasionally using a little wine or brandy with the sugar. The cost of a dish large enough for half a dozen persons will be covered by ten cents, unless it is made when apples are scarce and dear.

Apple Snow.—Make this dish when eggs are cheap. Pare and core a quart of apples, (cost five cents,) stew them to a pulp with just water enough to moisten them, rub them through a seive, and sweeten them to taste. Beat the whites of six eggs, (cost six cents,) with two [Pg 68] tablespoonfuls of powdered sugar, to a stiff froth; beat the apple-pulp to a froth; mix the egg and apple together very lightly, turning the bowl of the spoon over and over instead of stirring it around; then beat them with an egg whisk until they look like snow. Pile the snow high in the centre of a dish, putting it in by the tablespoonful, and taking care not to break it down; in the top of the heap of snow put a fresh flower or a green sprig; and if you have any currant jelly, lay a few bits around the base. The effect of the dish is very pretty, and it can be made for about fifteen cents.

Apple Cakes.—Pare, core, and slice a quart of apples, (price five cents,) stew them with half their weight in sugar, (about one pound, cost about twelve cents,) the grated rind and juice of a lemon, (cost two cents,) one ounce of batter, (cost two cents,) and a very little grated nutmeg. When they are tender beat them with an egg whisk until they are light, drop them by the dessert-spoonful on buttered paper laid on a baking sheet, and bake them in a cool oven until they are firm, which will be in about fifteen minutes. When they are cool put them in a tin box until wanted for use. The cost will be about twenty cents.

Cherry Cheese.—Put into a stone jar a pound of sound, ripe cherries, with the stones removed, (cost about ten cents;) cover the jar closely, set it in a saucepan half full of boiling water, and simmer it gently until the cherries are tender; then take up the fruit, weigh it, put it into a preserving kettle with half a pound of finely sifted sugar, (cost about eight cents), to every pound of fruit; add a dozen cherry kernels with the skins removed by scalding, and rubbing in a clean cloth, put the kettle over a slow fire, and boil, stirring occasionally, until the fruit is quite dry and clear. Meantime rinse out some shallow jars with brandy, and when the fruit is done put it into them, pressing it down tightly; pour a very little brandy over the top, lay a little paper on each, then fit on the covers of the jars closely, and keep in a dry, cool place. The above quantity will cost about twenty-five cents.

Candied Cherries.—Choose a pound of perfectly sound, ripe cherries, (cost ten cents,) with the stalks and an occasional leaf attached, wipe them with a clean, dry, soft cloth; dip the leaves and stems, but not the fruit, into boiling vinegar, and set them with the cherries upward, in a card-board perforated with holes to admit the stems, until the [Pg 69] vinegar dries. Meantime boil a pound of loaf sugar, (cost about fifteen cents), with a teaspoonful of cold water, using a thick porcelain-lined saucepan or copper sugar boiler; skim until perfectly clear, and test in the following way: Dip the thumb and forefinger into cold water and then quickly into the boiling sugar, withdrawing it instantly; press the fingers together, and then draw them apart; if the sugar forms a little thread between them it is ready to use, if it does not, boil a few minutes longer and test again. When it is ready dip the leaves and branches into it, and dry them in the card board frame as directed above. Keep the sugar at the boiling point, and as soon as it forms a clear brittle thread between the fingers, when tested as above, dip the entire fruit into it, moving the cherries around so that the sugar completely covers them, and dry them, placed as above in the card board frame, in the mouth of a cool oven.

Currant Salad.—Remove the stems from half a pound each of red and white currants, (cost ten cents,) and pile them in regular layers high in the centre of a shallow glass dish, sifting a little powdered sugar between each layer; the sugar will cost two or three cents. A gill of cream, (cost five cents,) may be poured over the top, if desired. The dish should be tastefully ornamented with green leaves, and the salad kept very cool until wanted for use. The cost of a nice dishful will be about eighteen cents.

Iced Currants.—Beat the white of one egg, (cost one cent,) to a stiff froth, mix it with three dessertspoonfuls of cold water, dip into it carefully some perfect bunches of ripe red and white currants, which can be bought in season for ten cents a pound; drain each bunch a moment and then dust it well with powdered sugar, lay each bunch carefully upon a large sheet of white paper, so that there is plenty of room between the bunches, and set them in a cool, airy place for five hours. The sugar will partly crystalize upon the fruit, and the effect will be very pretty. The cost of a good sized dish will be about fifteen cents.

Compôte of Damsons.—Wipe one quart of sound, ripe damsons, (cost ten cents,) with a clean, dry cloth, drop them, one by one into the following syrup: make a syrup by melting one pound of loaf sugar (cost fifteen cents,) with one pint of water, stir in the white of an egg, (cost one cent,) and boil the syrup fifteen minutes, skimming it clear. Simmer the plums in this syrup until they are tender, about five [Pg 70] minutes, but do not let them break; take the plums up carefully and arrange them in a heap on a shallow dish, letting the syrup boil about ten minutes, until it is quite thick; then remove it from the fire, cool it a little, and pour it over the plums. The dish will cost about twenty-five cents.

Stuffed Dates.—Remove the stones from a pound of fine dates, (cost ten cents,) by cutting them open at one side; remove the shells and skins from half a pound of almonds, (cost ten cents;) the skins can easily be rubbed off by first pouring boiling water on the almond kernels; replace the date-stones with the almonds, and arrange the dates neatly on a shallow dish; dust a little powdered sugar over them, and keep them in a cool, dry place till ready to use. The dish will cost twenty-three cents.

Stewed Figs.—Dissolve four ounces of powdered sugar, (cost three cents,) in one pint of cold water, and flavor with a few drops of any essence preferred; put into it a pound of nice figs, (cost ten cents,) heat slowly, and stew gently for about two hours, or until the figs are tender. Eat hot with a dish of boiled rice, or serve cold. The cost will be less than fifteen cents.

Compôte of Gooseberries.—Choose a quart of large, sound, ripe, green gooseberries, (cost ten cents,) remove the stems and tops, throw them into boiling water for two minutes; drain them, let them lay three minutes in cold water containing a tablespoonful of vinegar, to restore their color, and then drain them quite dry. Meantime make a thick syrup by boiling one pound of sugar, (cost twelve cents,) with one pint of water; as soon as the syrup has boiled about ten minutes, put in the gooseberries and boil them gently until just tender, about ten minutes. Then pour both fruit and syrup into an earthen or glass dish; cool, and use. The dish will cost less than twenty-five cents.

Gooseberry Cheese.—Remove the tops and stalks from two quarts of ripe, red gooseberries, (cost twenty cents,) put them in a moderate oven till soft enough to rub through a seive; then add to them one-fourth their weight of sugar, set them over the fire to boil gently for half an hour, stirring them constantly, and skimming till clear; then put by the tablespoonful on plates, and dry in the mouth of a cool oven. Pack, when quite cool, in a tight box, between sheets of white wrapping paper. The above quantity will cost about twenty-five cents.

Gooseberry Fool.—Remove tops and stalks from two quarts of goose [Pg 71] berries, boil them with three quarters of their weight in sugar, and half a pint of cold water, until soft enough to pulp through a sieve; then mix the pulp with a pint of milk, or cream, if a more expensive dish is desired, and put into an earthen or glass dish to cool; serve cold. The above quantity will cost about twenty-five cents.

Grape Jelly.—Dissolve one ounce of gelatine, (cost eight cents,) in half a pint of cold water. Break one pound and a half of grapes, (cost ten cents,) in an earthen bowl with a wooden spoon; strain the juice without pressing the grapes, through clean muslin, three times; put the juice into a preserve kettle with half a pound of loaf sugar, (cost eight cents,) and the dissolved isinglass, and boil it ten minutes; rub a jelly mold with pure salad oil; add two tablespoonfuls of brandy, (cost three cents,) to the jelly; pour it into the mould, and cool until the jelly sets firm. The above ingredients will make about a pint and a half of jelly, and will usually cost about twenty-five cents, for the above estimate is rather more than the average cost.

Green Gage Compôte.—Remove the skin from a quart of very ripe green gages, (cost fifteen cents,) put them in a glass dish, sprinkle them over with a pound of powdered sugar, (cost ten cents,) and let them stand in a cool place four hours, until a nice syrup has been formed. The dish is delicious, and costs about twenty-five cents.

Pine Apple Julep.—Pare and slice a very ripe pine apple, which in season will cost about ten cents; lay it in a glass dish; pour over it the juice of one orange, (cost two cents,) the juice of one lemon, (cost two cents,) a gill of any fruit syrup, (cost about five cents,) and two tablespoonfuls of rum, (cost three cents;) sprinkle it with a little powdered sugar, cool it on the ice, and serve it cold. It will cost about twenty-five cents.

Lemon Snow.—Soak one ounce of gelatine, (cost eight cents,) in one pint of cold water for half an hour; peel the yellow rind from three lemons, (cost six cents,) and squeeze and strain their juice; put the rind and juice of the lemons into a saucepan with eight ounces of loaf sugar, (cost eight cents,) and stir until the sugar and isinglass are quite dissolved; pour it into a bowl, and let it cool, and begin to grow firm. Then add the whites of three eggs, (cost three cents,) and beat to a stiff froth. Pile by the tablespoonful high in the centre of a glass dish. It is pretty and delicious, and costs only about twenty-five cents.

Melon Compôte.—Make a syrup by boiling one pound of sugar, [Pg 72] (cost ten cents,) with half a pint of water. Pare and slice a spicy musk melon, (cost five cents,) and put it into the syrup with a little wine, (cost five cents.) Boil gently for ten minutes, take up the melon in a glass dish, cool the syrup a little, and pour it over the melon. Serve the *compôte* cold; it is delicious, and costs only about twenty-five cents.

Orange Salad.—Peel six oranges, (cost twelve cents,) slice them, place them in rings in a glass dish, sprinkle them with three ounces of powdered sugar, (cost two cents,) pour over them a little wine and brandy, and let them stand over night in a cool place. A good dish full will cost about twenty cents.

Orange and Apple Compôte.—Pare and slice very thin three oranges, (cost six cents,) and three apples, (cost three cents,) removing the seeds from both: lay the slices in rings in a glass dish, cover, with the following syrup, and cool. Boil the orange peel in half a pint of water, with four ounces of sugar, (cost four cents,) until the syrup is clear; add a tablespoonful of brandy to it, cool it a little, and pour it over the sliced fruit. The dish is very nice when iced, and costs about fifteen cents.

Peach Salad.—Pare and quarter a quart of ripe peaches, (cost ten cents,) lay them in a heap in a shallow glass dish; squeeze over them the juice of an orange, (cost two cents,) and sprinkle them with powdered sugar, (cost two cents.) Put them on the ice to get very cold. A large dishful can be prepared for fifteen cents.

Cold Compôte of Pears.—Peel and slice thin a quart of Bartlett pears, (cost fifteen cents,) lay them in a glass dish, pour over them a little wine, and sprinkle them plentifully with powdered sugar. Let them stand in a cool place for an hour before using them. A nice dish will cost less than twenty cents.

Stewed Prunelles.—Put a pound of prunelles, (cost fifteen cents,) in enough boiling water to cover them, and stew them gently for one hour. Take them up with a skimmer, strain their juice, return it to the fire with four ounces of loaf sugar, (cost four cents,) the yellow rind and juice of one lemon, (cost two cents,) and a glass of wine; skim until clear, add the prunelles, and stew again for one hour. Take up the prunelles in a glass dish, cool the syrup a little, and strain it over them. Cool before using. The dish can be made for about twenty-five cents.

Quince Cakes.—Wash some quinces, boil them in enough water to [Pg 73] cover them, until they are tender enough to rub through a seive; to each quart add a pound and a half of loaf sugar, place the mixture over the fire, and heat to the boiling point, stirring it constantly, but do not let it boil. Oil some plates, spread the quince upon them, and dry it in the mouth of a cool oven. Then cut it in cakes, pack it in a tin box, between layers of white wrapping paper, when it is thoroughly cold, and keep it in a cool, dry place. A good dishful can be made for twenty-five cents.

Quince Snow.—Boil some nice quinces until tender, peel them, rub them through a sieve with a wooden spoon, and add to each pound a pound of powdered sugar, (cost ten cents,) and the whites of three eggs, (cost three cents.) Beat with an egg whisk to a stiff froth and pile by the tablespoonful in the centre of a shallow glass dish. A nice dishful can be made for about twenty-five cents.

Iced Raspberries.—Beat the white of one egg, (cost one cent,) with two tablespoonfuls of cold water; pick over a quart of fine ripe raspberries, (cost ten cents,) dip them one by one into the egg, and roll them in powdered sugar; lay them on white paper spread on a baking sheet, so that they do not touch, and dry them in a cold, dry place, sifting a little more sugar over them, if they seem to grow moist. When the berries are in season, twenty-five cents will cover the cost of a large dish.

Raspberry Salad.—Pick over a quart of ripe raspberries, (cost ten cents,) pile them high in the centre of a glass dish, pour over them a glass of wine, (cost five cents,) dust them with an ounce of powdered sugar, (cost one cent,) and keep on the ice till used. A good dishful can be made for about twenty cents.

Compôte of Strawberries.—Carefully pick over a quart of ripe strawberries, (cost ten cents;) put them in an earthen dish, pour over them a syrup made by boiling quarter of a pound of sugar, (cost four cents,) with one gill of water, for ten minutes; let the berries stand in this syrup for one hour; then drain them and pile them in a heap in a shallow glass dish; add to the syrup the juice of one orange, (cost two cents,) or a glass of wine; boil it up and cool it a little, and strain it over the berries; cool and use. This delicious dish costs about fifteen cents.

Strawberry Drops.—Rub some ripe strawberries through a fine seive with a wooden spoon; add two ounces of this juice to half a [Pg 74] pound of powdered sugar, (cost five cents,) put the mixture into a saucepan and stir it over the fire until it begins to simmer; remove it from the fire, and stir it briskly for five minutes, oil some paper, lay it on a baking sheet, drop the strawberries on it by the salt-spoonful, dry them in the mouth of a cool oven. Keep them between layers of white paper in a cool place. A good supply can be made for twenty-five cents.

Compôte of Mixed Fruit.—Boil half a pound of loaf sugar, (cost eight cents,) with one gill of cold water for ten minutes; pick over half a pound of red currants, (cost five cents,) and a pint of raspberries, (cost five cents,) and simmer them in the syrup for ten minutes. Take up the fruit on a glass dish, cool the syrup a little and pour it over the fruit. The dish will cost less than twenty cents.

Fruit Juice.—Rub ripe fruit through a seive, with a wooden spoon, and then strain it free from skins and seeds; to every pound add quarter of a pound of loaf sugar; mix well; put into wide-mouthed glass bottles, and set them in a pan with cold water reaching to the necks of the bottles. Set the pan over the fire and let the water come to a boil; remove the pan and let the bottles stand in the water until they are quite cold. Then cork them tightly, and seal them with wax or resin.

Keep them in a cool, dry place. This juice added to ice-water, and sweetened to taste, makes a delicious sherbet.

THE END.

INDEX.

	COST.	PAGE.
À la Mode Beef, with potatoes,	35	57
Apple Dumplings,	15	64
Bacon and Apple Roly-poly, with Vegetables,	25	48
Baked Apple Dumplings,	18	64
Baked Heart,	25	47
Baked Pig's Head,	22	50
Barley Water,	2	23
Batter for frying,	4	52
Batter for frying,	5	59
Beans and Bacon,	10	40
Beans, Baked,	10	39
Beans, Fried,	10	40
Bean Soup,	10	33
Beans, Stewed,	10	40
Beef *À la Mode*,	35	57
Beef Broth, with Dumplings,	25	37
Beef Patties,	30	58
Beef Pie,	25	46
Beer,	--	22
Beer for Nursing Women,	--	23
Biscuits, Sweet,	17	66
Biscuits, Tea,	6	27
Blanquette of Veal, with Potatoes,	30	61
Brain and Liver Pudding,	15	56
Bread per 8 lbs,	24	25
Bread Pudding,	15	63
Breakfast Rolls,	6	27

Broth, Beef,	25	37
Broth, Chicken,	5	53
Broth, Mutton,	17	36
Broth, Scotch,	10	32
Broth, Veal,	13	36
Broth, White,	25	36
Cake, Caraway,	24	66
Cake, Rice,	15	65
Cake, Rock,	22	65
Chicken Broth,	5	53
Chicken, Fried,	35	52
Cheese Pudding,	12	41
Chocolate,	--	21
Chowder,	20	35
Cocoa,	--	21
Cocoa, per quart,	6	22
Codfish Steaks and Potatoes,	20	45
Coffee,	--	21
Coffee, per quart,	6	22
College Pudding,	18	62
Cooking,	--	16
Cream Rice Pudding,	15	63
Cream Sauce,	6	62
Cream Soup, with Macaroni,	25	37
Croquettes, Rice,	20	65
Cup Custards,	12	63
Dumplings, Apple,	15	64
Dumplings, Apple, Baked,	18	64
Dumpling, Fruit,	20	64
Dumpling, Gammon,	15	48
Dumplings, Lemon,	23	64

Dumplings, Norfolk,	7	37
Dumplings, Suet,	5	53
Dessert,	10 to 25	67
Fish Chowder,	20	35
Fish Pudding,	25	44
Fish Soup,	20	35
Fish and Potato Pie,	25	44
Fish and Potato Pudding,	15	45
Forcemeat for Poultry,	10	51
Forcemeat for Veal,	5	60
Fowl, Roast,	38	51
Fruit Dumplings,	20	64
Fruit Tarts,	12	65
Gammon Dumpling,	15	48
German Potatoes,	10	55
Gingerbread, Soft,	10	66
Half-pay Pudding,	20	63
Hasty Pudding,	4	42
Indian Bread,	5	42
Indian Cakes,	5	42
Indian Pudding, Baked,	15	43
Indian Pudding, Boiled,	10	42
Irish Stew,	23	49
Johnny Cake,	5	42
Kidneys, Broiled, with potatoes,	20	56
Kidneys, Pigs',	10	47
Kidney Pudding,	30	47
Kidney, Stewed, with potatoes,	25	47
Kromeskys,	20	58
Lamb, *Epigramme*, with Broth and Rice,	20	59
Lemon Dumplings,	23	64

Lentils, Boiled,	14	41
Lentils, Fried,	10	41
Lentil Soup,	10	33
Lentils, Stewed,	10	41
Lime Water,	--	24
Liver Polenta,	12	57
Macaroni, Farmers' Style,	10	28
Macaroni, Milanaise Style,	13	28
Macaroni, with Broth,	10	28
Macaroni, with Cheese,	12	28
Macaroni, with Tomato Sauce,	18	29
Macaroni, with White Sauce,	10	28
Maize,	--	41
Marketing,	--	10
Measuring,	--	19
Meat Brewis,	5	38
Meat Patties, with Potatoes,	30	58
Milk,	--	23
Mutton Boiled, with Turnips and Potatoes,	17	58
Mutton Broth, with Vegetables,	17	36
Mutton Kromeskys, with Potatoes,	20	58
Mutton and Onions,	30	48
Mutton *rechauffée*, with Potatoes,	15	58
Norfolk Dumplings,	7	37
New York Cooking School Fricassee,	43	53
Oatmeal and Peas,	13	38
Onion Soup,	10	34
Oxtail Stew, with Bread,	22	46
Patties, Beef, with Potatoes,	30	58
Patties, Veal and Ham,	30	61
Peas and Bacon,	25	39

Peas and Onions,	10	39
Peas, Baked,	10	39
Peas Pudding,	10	39
Pea Soup,	10	33
Pea Soup, thick,	6	33
Pickled Shad, with bread,	20	54
Pigs' Head, Baked,	22	50
Polenta,	5	41
Polenta, Liver,	12	57
Pork and Onions,	20	49
Pork Chops, with Potatoes,	25	55
Pork Pie,	20	54
Pork, Roast, with Apples,	27	55
Potato Bread, per 8 lbs,	24	26
Potatoes, German,	10	55
Pudding, Brain and Liver,	15	56
Pudding, Bread,	15	63
Pudding, Cheese,	12	41
Pudding, College,	18	62
Pudding, Cream Rice,	15	63
Pudding, Fish and Potato,	25	45
Pudding, Half-pay,	20	63
Pudding, Hasty,	4	42
Pudding, Kidney,	30	47
Pudding, Peas,	10	39
Pudding, Swiss, with Sauce,	20	62
Pulled Bread,	3	26
Rabbit Curry,	28	53
Rabbit Pie,	30	54
Red Herrings and Potatoes, with Bread,	22	45
Rice, Boiled,	7	30

Rice Bread, per 8 lbs,	25	26
Rice Cake,	15	65
Rice Croquettes,	20	65
Rice, Japanese Style,	10	30
Rice, Milanaise Style,	10	30
Rice Milk,	15	35
Rice Panada,	12	30
Roast Fowl,	38	51
Roast Pork, with Apples,	27	55
Roast Veal, with Potatoes,	30	60
Rock Cakes,	22	65
Rolls, Breakfast,	6	27
Salt, Celery,	--	19
Salt, Spice,	--	19
Sauce, Cream,	6	62
Sauce, Table, per pint,	6	19
Sauce, Tomato,	10	29
Sausage, Stewed,	25	55
Scotch Broth, without Meat,	10	32
Seasoning,	--	18
Soft Gingerbread,	10	66
Swiss Pudding, with Sauce,	20	62
Shad, Pickled,	20	54
Sheep's Head Stew,	25	46
Sheep's Haslet,	17	49
Soup, Bean,	10	33
Soup, Cream,	25	37
Soup, Fish,	20	35
Soup, Lentil,	10	33
Soup, Onion,	10	34
Soup, Pea,	10	33

Soup, Spinach,	15	34
Soup, Thick Pea,	6	33
Soup, Vegetable,	20	34
Spinach Soup,	15	34
Stuffing for Poultry,	10	51
Stuffing for Veal,	5	60
Suet Dumplings,	5	53
Sweet Biscuits,	17	65
Table Sauce, per pint,	6	19
Tarts, Fruit,	12	65
Tea,	--	21
Tea Biscuit,	6	27
Tea, per quart,	3	22
Tincture Lemon,	--	19
Tincture Orange,	--	19
Tincture Vanilla,	--	19
Tomato Sauce,	10	29
Tripe, Curry and Rice,	27	56
White Broth, with Macaroni,	25	36
Veal and Ham Patties,	30	61
Veal and Rice,	20	49
Veal, *Blanquette*, with Potatoes,	30	61
Veal Broth, with Vegetables,	13	36
Veal, Roast, with Potatoes,	30	60
Vegetable Soup and Bacon,	20	34
Vegetable Porridge,	15	35

[Pg 81]

NOW READY.

A Household Treasure,

EXPLAINING

The System of Economical Cookery taught in the New York Cooking School.

MISS CORSON'S

COOKING SCHOOL TEXT-BOOK

AND

HOUSEKEEPERS' GUIDE

TO

Cookery and Kitchen Management.

12mo, Cloth, price, by mail, $1.25.

"HOW WELL CAN WE LIVE IF WE ARE MODERATELY POOR?"

The economical housewife will find this question answered in MISS CORSON'S

COOKING MANUAL.

18mo, Enamelled Cloth. Price, by mail, 50 cents.

ADDRESS,

ORANGE JUDD COMPANY,

245 Broadway, New York;

OR,

NEW YORK COOKING SCHOOL OFFICE,

35 East 17th Street, New York. [Pg 82]

IN PREPARATION,

and will be published by

ORANGE JUDD COMPANY,

an entirely new and most valuable work entitled

Good Cooking for Everybody,

By Miss JULIET CORSON.

A book that will be wanted by Every Housekeeper.

The American Agriculturist

FOR THE

FARM, GARDEN, AND HOUSEHOLD.

Established in 1842.

The Best and Cheapest Agricultural Journal in the World.

Terms, which include postage *pre-paid* by the Publishers: $1.50 per annum, in advance; 3 copies for $4; 4 copies for $5; 5 copies for $6; 6 copies for $7; 7 copies for $8; 10 or more copies, only $1 each. Single Numbers, 15 cents.

The Amerikanischer Agriculturist.

The only purely Agricultural German paper in the United States, and the best in the world. It contains all of the principal matter of the English Edition, together with special departments for German cultivators, prepared by writers trained for the work. Terms same as for the "American Agriculturist."

ORANGE JUDD COMPANY, 245 BROADWAY, NEW YORK.

www.ingramcontent.com/pod-product-compliance
Lightning Source LLC
Chambersburg PA
CBHW030443220526
45464CB00006B/2392